Radiocarbon Dating
An Archaeological Perspective

Radiocarbon Dating
An Archaeological Perspective

R. E. Taylor

Department of Anthropology
Institute of Geophysics and Planetary Physics
University of California, Riverside
Riverside, California

1987

ACADEMIC PRESS, INC.

Harcourt Brace Jovanovich, Publishers
Orlando   San Diego   New York   Austin
Boston   London   Sydney   Tokyo   Toronto

ACADEMIC PRESS, INC.
Orlando, Florida 32887

United Kingdom Edition published by
ACADEMIC PRESS INC. (LONDON) LTD.
24–28 Oval Road, London NW1 7DX

Library of Congress Cataloging in Publication Data

Taylor, R. E. (Royal Ervin), Date
    Radiocarbon dating.

    Bibliography: p.
    Includes index.
    1. Archaeological dating.   2. Radiocarbon dating.
I. Title.
CC78.T39   1987   930.1'028'5   86-22222
ISBN 0—12—684860—2 (alk. paper)

PRINTED IN THE UNITED STATES OF AMERICA

87 88 89 90   9 8 7 6 5 4 3 2 1

To Marilynn, Gregory, and Karen

Contents

PREFACE

In December 1960, the Nobel Prize for Chemistry was bestowed upon Willard Frank Libby (1909–80) for, in the words of the citation, "his method to use Carbon-14 for age determinations in archaeology, geology, geophysics, and other branches of science." The chairman of the Nobel Committee for Chemistry for that year reported that one of the scientists who suggested Libby as a candidate for the Nobel laureate characterized his work in these terms: "Seldom has a single discovery in chemistry had such an impact on the thinking of so many fields of human endeavor. Seldom has a single discovery generated such wide public interest" (Nobel Foundation, 1964:587–588).

The application of the ^{14}C method to archaeological materials is generally considered to be a watershed in the history of archaeology and, in particular, in prehistoric studies. The impact of ^{14}C age estimates on a broad spectrum of important problems and issues has been and continues to be crucial. Glyn Daniel, retired holder of the archaeology chair at Cambridge, has gone so far as to equate the development of the ^{14}C method of the 20th century with the discovery of the antiquity of the human species in the 19th century. He insisted that the method represents the great revolution in 20th-century studies of prehistory (Daniel, 1967:266; 1959:79–80). The current Cambridge professor Colin Renfrew is known widely for his volume *Before Civilization: The Radiocarbon Revolution and Pre-historic Europe* (Renfrew, 1973). One writer has even suggested that the impact of ^{14}C dating on archaeology can be legitimately compared to the influence of the discovery of the periodicity of the elements in chemistry (Keisch, 1972:25). While such a comparison may be somewhat forced, it is certainly correct that, on a worldwide basis, ^{14}C data provide the central core around which the late Pleistocene and Holocene prehistoric time scales have been built (cf. Aitken, 1974:27; Clark, 1961, 1969).

It appears that Libby's first formal public lecture describing the ^{14}C method was made before a group of archaeologists and anthropologists in January of 1948 in New York City. Since then, the technique has been extensively reviewed and discussed. The founding statement, of course, is Libby's own work *Radiocarbon Dating*. The second and final edition of this volume was issued in 1955. The 1965 paperback second edition included only a small number of addendum notes. Yet, because of the number of disciplines that benefit from ^{14}C data, the literature dealing with the method has become voluminous and specialized. International conferences on radiocarbon dating have provided opportunities for periodic appraisals of advances in ^{14}C technology and its applications in many disciplines.

Over the years, a number of discussions that focused on the basis of ^{14}C dating and its applications in archaeology have appeared separately or in textbook or symposium volumes. These have included Aitken (1974), Barker (1970), Berger (1970b), Broecker and Kulp (1956), Fleming (1976), Gillespie (1984), Michels (1973), Mook and Streurman (1983), Ogden (1977), Olsson (1968), Polach (1976), Polach and Golson (1966), Ralph (1971), Sheppard (n.d.), Taylor (1978), Tite (1972:76–90), and Willis (1969). The purpose of this volume is to provide an introductory overview of the radiocarbon dating method in terms of issues and applications in archaeology, particularly in the context of the development of various points of view in the application of the technique. Because of this focus, references to the literature cited in the text have been used in some cases to illustrate the history of various aspects of ^{14}C geochemistry or geophysics or applications of ^{14}C data to a specific body of archaeological or geological data.

In 1969 at the *Nobel Symposium on Radiocarbon Variations and Absolute Chronology*, Professor Torqny Säve-Söderbergh, quoting Professor J. O. Brew of Harvard, suggested that an attitude among certain archaeologists toward radiocarbon determinations could be summarized in these words:

> If a C14 date supports our theories, we put it in the main text. If it does not entirely contradict them, we put it in a foot-note. And if it is completely 'out of date' we just drop it (Säve-Söderbergh and Olsson 1970:35).

The need to even contemplate this approach has been "out of date" among most archaeologists since the very beginning of the ^{14}C method. There has been, however, a recognition of the dangers of uncritically accepting ^{14}C age estimates without careful analyses, on a case-by-case basis, of the various factors that can affect accuracy and precision. This volume has been written to provide a review of some of the major advances and accomplishments of the ^{14}C method from an archaeological perspective. It is also intended to provide an introduction to some of the problems and issues involved in the use of ^{14}C data in archaeological studies.

R. E. Taylor
Loma Linda, California

ACKNOWLEDGMENTS

The author would like to acknowledge the assistance of the late Willard F. Libby in the preparation of both this volume and an earlier chapter that appeared in *Archaeological Chemistry II*, published as part of the *Advances in Chemistry* series of the American Chemical Society (Taylor, 1978). The origin of both of these works lies in the opportunity given to a first-year graduate student in archaeology/anthropology at the University of California, Los Angeles (UCLA), to serve as a research assistant in the Isotope Laboratory of the UCLA Institute of Geophysics and Planetary Physics. Individuals who have commented on various sections of this volume include Ernest C. Anderson, Paul E. Damon, C. Wesley Ferguson, Herbert Haas, Henry Polach, Robert Stuckenrath, Minze Stuiver, Irwin P. Ting, and Philip J. Wilke. A. J. T. Jull has been very helpful in the preparation of the section of ^{14}C accelerator mass spectrometry. Paul E. Damon kindly provided a manuscript copy of a very informative paper titled "The History of the Calibration of Radiocarbon Dates by Dendrochronology" and also provided helpful comments. Any errors or misunderstanding of information provided from any source is entirely the responsibility of the author.

Chapter 6, which deals with the history of ^{14}C dating, could not have been written without information and comments provided by Ernest C. Anderson, James R. Arnold, Frederick Johnson, and Leona Marshall Libby. A shortened version of portions of Chapter 6, which focused on materials published in *American Antiquity*, has previously appeared (Taylor, 1985a). Photographs of the Chicago equipment and access to the files of the Chicago Laboratory (now housed in the Isotope Laboratory, UCLA) were made available by Rainer Berger. Leona Marshall Libby kindly allowed access to an unpublished manuscript (L. M. Libby, n.d.) that contains important information pertaining to aspects of the work on ^{14}C dating at the University of California, Berkeley, and the University of Chicago. Permission to quote from a transcript of a 1979 interview with W. F. Libby was obtained from the Center for History of Physics, American Institute of Physics, New York. Gregory Marlowe, who conducted this interview, has provided additional comments and information. The translation of the Swedish text of the Nobel award to W. F. Libby was carried out by Gorman Bjorch, University of Stockholm, and Tord Ganelius, Secretary of the Royal Swedish Academy of Science. Anne Kernan and Allen D. Zych of the University of California, Riverside (UCR), Physics Department were very helpful in designing a representation of the carbon atom that appears on the cover of this book.

The author is indebted to Peter J. Slota, Jr., for his invaluable contribution to the work done at the UCR Radiocarbon Laboratory for more than a decade as well as to the dedicated assistance and involvement of Paul Ennis, Louis A. Payen, and Christine A. Prior. Others who have made important contributions to the UCR laboratory include Brooke Arkush, Jeanne Binning, Bradley Lowman, Edward Plummer, Jeffrey Simmons, and Donald K. Sullivan. Special thanks should also be expressed to Linda Bobbitt and Diana Deporto who prepared many of the figures included in this volume.

CHAPTER 1

ELEMENTS OF THE RADIOCARBON METHOD

1.1 BASIC PRINCIPLES

Carbon-containing compounds are widely distributed in many forms throughout the earth's diverse environments. These materials are cycled through the various carbon reservoirs on different time scales primarily by solar energy. This process includes the operation of two major interacting systems: (i) a photosynthetic cycle involving the fixation of atmospheric carbon dioxide in plant materials, an incorporation of a small portion of this in animal tissue, and subsequent decomposition and (ii) the cycling in the oceans, atmosphere, and major lake systems of various chemical species (carbon dioxide–carbonate–bicarbonate). Various geologic processes including the deposition of carbonates in sediments and volcanic activity are also involved in the operation of the carbon cycle (Craig, 1953, 1957).

Carbon has three naturally occurring isotopes, two of which are stable (^{12}C, ^{13}C) and one (^{14}C) which is unstable or radioactive.[1] The natural

---

[1] Over the years, various symbols (^{14}C, C^{14}, C-14, Carbon-14) have been used to designate radiocarbon. The current international convention is to use ^{14}C as the standard abbreviation.

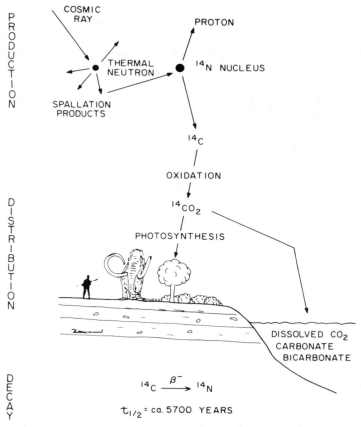

Figure 1.1 Basis of radiocarbon method: production, distribution and decay of ^{14}C. [After Taylor (1985b).]

radioactive isotope of carbon, or radiocarbon, decays with a half-life of about 5700 years. The basis of the ^{14}C dating method can be simply illustrated, as in Fig. 1.1, in terms of the production, distribution, and decay of ^{14}C. The natural production of ^{14}C is a secondary effect of cosmic-ray bombardment in the upper atmosphere. Following production, it is oxidized to form ^{14}CO$_2$. In this form, ^{14}C is distributed throughout the earth's atmosphere. Most of it is absorbed in the oceans, while a small percentage becomes part of the terrestrial biosphere. Metabolic processes maintain the ^{14}C content of living organisms in equilibrium with atmospheric ^{14}C. Once metabolic processes cease—as at the death of an animal or plant— the amount of ^{14}C will begin to decrease by decay at a rate measured by the ^{14}C half-life.

The *radiocarbon age* of a given sample is based on measurement of residual ^{14}C content. While the theoretical maximum age range using accelerator mass spectrometry (AMS) for direct counting of ^{14}C (Section 4.5) may eventually approach 100,000 years (cf. Muller, 1977; Libby, W. F., 1979), current experimental conditions reduce this to between 40,000 and 60,000 years. With conventional decay counting (Section 4.4), practical limitations imposed by sample sizes generally available from archaeological contexts and the problem of removing contamination reduce the maximum range in the vast majority of samples to between 40,000 and 50,000 years. Under special circumstances and by employing relatively large sample sizes usually not available to archaeologists, the maximum range can be extended to about 60,000 years (e.g., Geyh, 1965; Stuiver et al., 1979). With isotopic enrichment—again employing relatively large amounts of sample material—ages up to 75,000 years have been reported on a small number of samples (Grootes et al., 1975; Stuiver et al., 1978; Erlenkeuser, 1979). Developments in AMS technology, combined with new approaches to the enrichment of the ^{14}C content in milligram size samples using laser technology, may in the future permit the practical extension of the ^{14}C time frame beyond 50,000 years on a routine basis for the typical archaeological sample, if the stringent requirements in the exclusion of contaminants both in field and laboratory contexts can be met (Hedges et al., 1980; Stuiver, 1978a).

1.2 GENERAL ASSUMPTIONS

For a radiocarbon age to be equivalent to its actual or calendar age, at a reasonable level of precision, several assumptions must hold within relatively narrow limits. They can briefly be summarized as follows (cf. Taylor, 1978:42):

1. The concentration of ^{14}C in each carbon reservoir has remained essentially constant over the ^{14}C time scale;
2. there has been complete and rapid mixing of ^{14}C throughout the various carbon reservoirs on a worldwide basis;
3. carbon isotope ratios in samples have not been altered except by ^{14}C decay since these sample materials ceased to be an active part of one of the carbon reservoirs (as at the death of an organism);
4. the half-life of ^{14}C is accurately known; and
5. natural levels of ^{14}C can be measured to appropriate levels of accuracy and precision.

An additional requirement for a critical utilization of the method is that there is a known relationship or association between the sample to be

analyzed and some specific event or phenomenon to be dated. The five
assumptions listed above are the subject of continuing study by radiocarbon specialists. A satisfactory fulfillment of the contextual requirement
is largely in the hands of the field archaeologist, geologist, or historical
specialist collecting and submitting samples for analyses.

1.3 CONVENTIONS

Radiocarbon age estimates are generally expressed in terms of a set of
widely accepted parameters that define a *conventional radiocarbon age*
(Stuiver and Polach, 1977). These conventions include (i) the use of 5568
(5570) years as the ^{14}C half-life even though the actual value is probably
closer to 5730 years[2] (Section 1.5); (ii) the direct or indirect use of a National Bureau of Standards (NBS)-distributed oxalic acid preparation (a
1957 "old" or 1977 "new" NBS oxalic acid) as a contemporary standard
to define the "zero" ^{14}C age in the terrestrial biosphere (Section 4.6); (iii)
the use of A.D. 1950 as the zero point from which to count ^{14}C time; (iv)
a correction or normalization of ^{14}C activity in all samples to a common
$\delta^{13}C$ value to account for fractionation effects (Sections 4.6 and 5.3.2);
and (v) an assumption that ^{14}C in all reservoirs has remained constant
over the ^{14}C time scale (Section 2.2). In addition, each ^{14}C determination
should be accompanied by an expression that provides an estimate of the
experimental or *analytical uncertainty*. Since statistical constraints are
usually the dominant component of the experimental uncertainty, this value
is sometimes informally referred to as the "statistical error" or the statistical deviation. This "±" term is suffixed to all appropriately documented ^{14}C age estimates (Sections 4.7 and 5.4). It is also customary that
a laboratory number and an appropriate primary bibliographic reference—
typically from a list of ^{14}C determinations measured by a given laboratory
published in the journal *Radiocarbon*—be associated with each ^{14}C age
citation.

For samples from some carbon reservoirs, the conventional contemporary standards may not define a zero ^{14}C age. A *reservoir corrected
radiocarbon age* can sometimes be calculated by documenting the apparent
age exhibited in control samples and correcting for the observed deviation
(Section 2.3). A *calibrated radiocarbon age* takes into consideration the
fact that ^{14}C activity in living organisms has not remained constant over

---

[2]A suggestion (Wigley and Muller, 1981:176) that the term *corrected radiocarbon age* be
used to designate a ^{14}C age estimate calculated with the 5730-yr ^{14}C half-life has not received
general acceptance in the radiocarbon community.

the ^{14}C time scale (Sections 2.2 and 5.5.2). Because of this, "radiocarbon years" as employed in a conventional ^{14}C age expression and "calendar years" for some time periods may not be of equal duration. Calibrated ^{14}C values for the last 8000 years are obtained by consulting tables or plots of relationships between ^{14}C and dendrochronologically (tree-ring)-dated samples to determine the amount of variation between them for different periods.

Radiocarbon age estimates are typically expressed in years B.P. (in some publications BP), i.e., years before the present. When it became customary to use A.D. 1950 as 0 B.P. (avoiding what some saw as a problem of having a B.P. ^{14}C value change with the passage of time), the meaning of B.P. ("before present" being the year in which the ^{14}C measurement had been made) was changed to mean "before physics," i.e., before A.D. 1950 (Flint and Deevey, 1962). The use of 1950 as the standard zero point was a choice reflecting the closest date to the publication of the first ^{14}C determinations in December 1949 (Arnold and Libby, 1949). Between 1963 and 1977, ^{14}C values were published in *Radiocarbon* in both their B.P. and A.D./B.C. forms—the B.C. or A.D. values being calculated by subtracting 1950 from B.P. values in excess of 1950 (B.C. expressions) or subtracting the B.P. values in the 0–1950 years range from 1950 (A.D. expressions). In 1977, following a recommendation of the 1976 International Radiocarbon Conference, *Radiocarbon* discontinued this practice. One laboratory suggested an *A.D./*B.C. nomenclature in which both the B.P. and a calibrated ^{14}C were expressed, e.g., P-1699 (P indicates the University of Pennsylvania laboratory), 2990 ± 50 (the conventional B.P. value), *1290 ± 50 B.C. [the calibrated ^{14}C value, using a calibration approach developed at the University of Pennsylvania Museum Applied Science Center for Archaeology (MASCA)] with the citation (Fishman *et al.*, 1977:201). This approach has been adopted by several laboratories (e.g., Nydal *et al.*, 1985) in some cases with minor changes in format (e.g., 1290 ± 50 B.C.*). Other laboratories calibrate their ^{14}C values using other calibration formats or schemes (e.g., Evin *et al.*, 1985). The editorial policy of some journals (e.g., *Quaternary Research*) also enjoins the use of an asterisk to denote calibrated ^{14}C values (i.e., 1290 B.C.*).

Earlier, another nomenclature had been adopted in the journal *Antiquity* as a means of distinguishing between conventional (uncalibrated) and calibrated ^{14}C values. It was suggested that "bp" (and "ad/bc") be employed to designate conventional ^{14}C values and "BP" and "AD/BC" be used to designate calibrated ^{14}C values (Daniel, 1972; cf. Suess and Strahm, 1970). Although this approach has been used in a number of publications, it did not receive international support from the radiocarbon community (cf. Berger and Suess, 1979:xii). The current practice of *Radiocarbon* is to publish only the conventional B.P. ^{14}C value as defined above. This decision was made in part because of the number of calibration schemes (Section 2.2) and a lack of consensus as to which scheme provides the most effective approach to estimating a calibrated ^{14}C value (Scott *et al.*, 1984). At the 12th (1985) International Radiocarbon Conference held in Trondheim, Norway (Stuiver and Kra, 1986), it was proposed that calibrated values be expressed with the notation Cal BP, Cal AD and Cal BC (Mook, 1986). The style of expression employed in this volume will refer to ^{14}C values primarily in terms of "^{14}C years B.P." Radiocarbon determinations interpreted in light of dendrochronologically determined equivalents are listed as calibrated values in B.C. or A.D. year units with the specific calibration approach explicitly identified.

1.4 PRODUCTION AND DISTRIBUTION OF NATURAL RADIOCARBON

The production of ^{14}C can take place as a result of several nuclear reactions. The dominant natural mechanism, however, involves the bombardment of ^{14}N by the cosmic-ray-produced neutrons in a relatively narrow range of energies (Lingenfelter, 1963; Castagnoli and Lal, 1980:154). These neutrons are produced in a spallation process as a result of the collision of the proton component of cosmic rays with gaseous components of the atmosphere. The term cosmic ray reflects the fact that the source of this extremely energetic radiation is beyond our solar system. A fraction of these neutrons escape into space, but a significant percentage remain. Through collisions with air molecules, they gradually lose energy and it is these slower or "thermal" neutrons that react with ^{14}N—nitrogen being the largest constituent of the atmosphere—to form ^{14}C and a proton (hydrogen nucleus). This reaction can be expressed as:

$$^{14}N + n = {}^{14}C + {}^{1}H \tag{1.1}$$

or

$$^{14}N(n, p){}^{14}C. \tag{1.2}$$

Other products of the reaction of neutrons with ^{14}N have been reported in laboratory experiments, but over 90% of these collisions will form ^{14}C. It might be noted that the great abundance of atmospheric nitrogen means that minor changes in nitrogen content in the atmosphere would have no effect on ^{14}C production, this being controlled only by the supply of neutrons. (There is no evidence for major changes in atmospheric nitrogen concentrations in geologically recent periods.) It has been generally accepted that within several minutes or, at most, several hours, virtually all ^{14}C atoms will have undergone oxidation to form carbon dioxide. However, the exact mechanism(s) involved are not well understood. It is as $^{14}CO_2$ that natural ^{14}C makes its initial entry into the terrestrial carbon cycle.

The possible *in situ* production of ^{14}C in plant structures (particularly wood) at relatively high altitudes by the direct action of cosmic-ray-produced neutrons (or as the result of neutrons produced by lightning strikes) has also been suggested by several investigators (e.g., Lal in Suess and Strahm, 1970:95; Baxter and Farmer, 1973b; Libby and Lukens, 1973; cf. Damon *et al.*, 1973). Based both on theoretical calculations and experimental data, it has been concluded that even under the most favorable circumstances, the effect would be negligible for wood even from high altitude locations certainly over the Holocene and probably over the last 50,000 years (Harkness and Burleigh, 1974; Radnell *et al.*, 1979; Cain and Suess, 1976). For materials with indicated ages in excess of 50,000–60,000 years that may contain higher amounts of nitrogen than are found in wood (e.g., bone), the effect might conceivably be noticeable. However, additional experiments are needed to determine if, even in such cases, the effect could be detected (Wand, 1981:254–270).

The rate of natural ^{14}C production in the atmosphere is not uniform with regard to geographical locality. Since the amount of ^{14}C produced is a function of neutron intensity, any variation in this value will directly affect the ^{14}C production rate. The cosmic-ray neutron intensity at the geomagnetic poles is about 5 times that at the geomagnetic equator. The principal reason for this effect is the shielding action of the earth's magnetic field which reflects incoming cosmic rays of lower energies back into space. This shielding effect is less at higher geomagnetic latitudes resulting in a latitude effect in the neutron intensity and consequently a variation in the rate of ^{14}C production. Fortunately, this gradient in ^{14}C concentration essentially disappears as individual $^{14}CO_2$ molecules are mixed by atmospheric turbulence, particularly in the lower portions of the earth's atmosphere. A slight latitude and altitude effect has been reported in tree rings of similar age growing at different locations on the earth's surface. However, this effect probably does not exceed, at most, about 0.5% or the equivalent of an age effect of about 40 years (Lerman *et al.*, 1970; Browman, 1981:258–260).

Once a ^{14}C-tagged CO_2 molecule reaches the surface of the earth, it enters into the carbon exchange cycle. Figure 1.2 provides a simplified representation of this system. For our purposes, the earth's carbon-containing environments have been partitioned into atmospheric, biospheric, hydrospheric (primarily oceans), and sedimentary reservoirs. The first three represent the dynamic portion of the system with characteristic

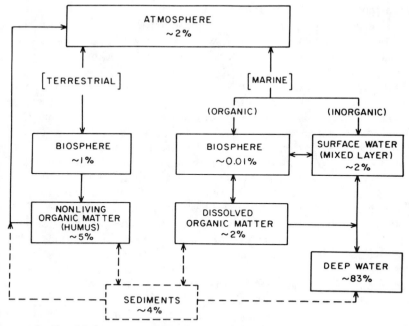

Figure 1.2 Simplified representation of carbon reservoirs. Freshwater values not included. Approximate percentage of ^{14}C in each major reservoir. [After Damon *et al.* (1983:250) and U. Siegenthaler, personal communication.]

exchange rates on the order of a few years for the atmosphere (and most of the biosphere) and up to about 1000 years for parts of the oceans. The rate of carbon exchange between the dynamic carbon reservoirs and the relatively inactive reservoirs (e.g., carbonates in sedimentary limestone deposits) is on a much longer scale and is generally thought not to enter into carbon reservoir calculations as far as the tracing of ^{14}C in dating applications is concerned (Broecker and Olson, 1959:112).

As we have noted, terrestrial plants fix ^{14}C into their cellular structures by photosynthesis. Other terrestrial organisms obtain it secondarily by ingestion of plant materials. The amount of ^{14}C in the terrestrial biosphere, however, is completely overshadowed by that contained in the oceans. About 85% of the ^{14}C on earth exists dissolved as CO_2, bicarbonate, or carbonate in ocean water (Fig. 1.2). All but a few percent of these inorganic carbonates are present in the deep ocean below the rapidly mixing layers of the surface waters.

The fact that most of the ^{14}C on earth resides in the deep ocean is of major importance to the success of the ^{14}C method of dating. This part of the carbon reservoir acts as a buffer that effectively dampens major fluctuations in terrestrial ^{14}C activity. This dampening effect means that any short term (<100 years) perturbation, even a major one, in ^{14}C activity from whatever cause will be modulated on the time scale of several hundred years. Because there is no evidence for secular or long term anomalies in biospheric ^{14}C activity exceeding 10% over the period of the Holocene (approximately the last 10,000 years), it is probable that the physical and biological constants controlling the behavior of the deep ocean with respect to ^{14}C concentrations (e.g., volume, temperature, pH, and biological productivity) have not changed drastically in that period. However, the nature of various parameters affecting the geochemistry of oceans during the late Pleistocene must be considered in evaluating ^{14}C determinations on materials with indicated ages in excess of about 10,000 years.

1.5   DECAY OF RADIOCARBON

Radiocarbon decays by emitting a beta particle to form ^{14}N and a neutrino. Beta particles are high-speed electrons ejected from nuclei and exhibit a spectrum of energies. The maximum energy of ^{14}C beta emissions is about 0.156 MeV (1 MeV $= 10^6$ electron volts), which classifies this isotope as a *weak* beta emitter. Because of its use in biochemical and biomedical research for over twenty-five years, the literature on ^{14}C as a biological tracer is extensive (e.g., Raaen *et al.*, 1968). However, the applications and detection technology developed to deal with relatively high tracer levels is only indirectly relevant to measurement of natural (low level) ^{14}C concentrations.

The fundamental constant that permits the conversion of ^{14}C concentration data into an apparent age is the half-life of ^{14}C. Chapter 6 briefly reviews the early history of the efforts to ascertain the ^{14}C half-life value. As we have noted, conventional radiocarbon dates are expressed with respect to the "Libby half-life" of 5568 ± 30 years. Redeterminations by a number of other researchers have led to a slight upward revision of the value (Olsson, 1968:207). Currently, the most widely quoted estimate is the "Cambridge half-life" of 5730 ± 40, which translates into about a 3% increase (Godwin, 1962b; cf. Hughes and Mann, 1964).

For middle and late Holocene ^{14}C values, the question of which half-life should be employed has lost much of its significance primarily because of the documentation of dendrochronologically determined relationships between ^{14}C ages and calendar ages (Section 2.2). Currently, researchers working with materials over about the last 8000 years can, if necessary, circumvent the problem of the actual half-life and calibrate ^{14}C age estimates directly. For ^{14}C values beyond the range of tree-ring calibration data, however, it is appropriate to be aware of the uncertainty in the half-life value. The use of the 5730 year half-life increases all calculated ^{14}C ages by 3%, i.e., a 35,000 year conventional ^{14}C age will be about 1000 years too young. This value is somewhat in excess of the typical statistical uncertainty for conventional counting systems. However, it is likely that fluctuations in ^{14}C concentrations and the effects of other as yet undocumented geophysical variations in the Pleistocene may produce effects greater than that represented simply by adjustments in the half-life value.

1.6 MAJOR BIBLIOGRAPHIC SOURCES

The primary bibliographic source for individual ^{14}C age determinations is "date lists" prepared by the various radiocarbon laboratories. The custom of publishing ^{14}C determinations in this form began with the first ^{14}C laboratory, that of Libby at the University of Chicago (Arnold and Libby, 1950, 1951). Each laboratory selected an alphabetic code abbreviation (e.g., C [= University of Chicago], BM [= British Museum], UCR [= University of California, Riverside], which, when prefixed to numerical values (e.g., C-530, BM-1320, UCR-1251, etc.), constitutes a unique laboratory number assigned to each sample. In some instances, samples in closely related series are assigned the same number but are distinguished by letter suffixes (e.g., BM-1320A, BM-1320B, BM-1320C, etc.). Each laboratory publishes its date lists in a sequentially numbered series (e.g., British Museum Radiocarbon Dates I, University of Arizona Radiocarbon Dates II, etc.)

By 1958, the increasing number of laboratories and the amount of space being devoted to the publication of ^{14}C date lists in the pages of *Science* resulted in a decision to begin a specialized journal (Deevey, 1984). It began in 1959 as the *Radiocarbon Supplement* to the *American Journal of Science,* an annual publication edited by Richard Foster Flint and Edward S. Deevey, Jr. of Yale University. By the third volume of this series (1961), the title was changed to *Radiocarbon.* In 1963, a Yale University archaeologist, Irving Rouse, was added to the staff. *Radiocarbon* expanded to two issues per year in 1958 and three in 1973. Although it was initially conceived as primarily a publication medium for date lists, over the last decade it has been expanded to include review articles, technical notes, and book reviews. Major policy statements reflecting international agreements dealing with the standard treatment of ^{14}C values are also published in *Radiocarbon*.

One of the unfortunate consequences of the great expansion in the number of ^{14}C laboratories over the last three decades has been the decreasing percentage of ^{14}C determinations that appear in date lists published in *Radiocarbon*. Accurate data are difficult to obtain, but one estimate suggests that, depending on the region, 25–50% of ^{14}C dates are not published in *Radiocarbon*. The total number of individual ^{14}C determinations probably now exceeds 100,000 with perhaps as many as 40% having been obtained on samples of archaeological interest (Taylor, 1981; Weinstein, 1984:297).

In addition to *Radiocarbon*, another important vehicle for the dissemination of research data in ^{14}C studies has been the periodic radiocarbon conferences. Table 1.1 lists the conferences generally included in this series. The first formal gatherings were held in 1954 in Europe (Copenhagen) and in the United States (Andover, Massachusetts). The first international radiocarbon conference was held in Cambridge, England, in the summer of 1955. Publication of the proceedings began with the 1965 Pullman, Washington, conference. Since that time, international radiocarbon conferences have been held on an approximately triennial basis. A second series of meetings concerned in part with ^{14}C dating was inaugurated as a result of the use of AMS for ^{14}C analysis (Section 4.5). The first AMS conference was held at the University of Rochester in the spring of 1978 and an approximately triennial schedule for these meetings has been maintained. (Papers on AMS ^{14}C research are also contributed at the international radiocarbon conferences.) The Oxford Radiocarbon Accelerator Unit, an AMS laboratory that focuses its attention on archaeological samples, has inauguated a practice of publishing results of its AMS ^{14}C analysis on archaeological samples in the journal *Archaeometry* (e.g., Gillespie *et al.*, 1984c, 1985; Gowlett *et al.*, 1986). These date lists are intended to serve as a formal interim means of publication for such data, but they will not be a substitute for final publication in *Radiocarbon*.

Any list of publications dealing with ^{14}C dating applications in archaeology reflects the extremely diverse nature of the subject matter. The interdisciplinary character of the ^{14}C field as it impacts archaeology insures that important works have been published in the physical science as well as in the anthropological and archaeological literature. Hilde Levi published the earliest bibliography of ^{14}C dating, a selected list of references from 1946 to 1954 (Levi, 1955a) followed by listings for the years 1956 and 1957 (Levi, 1957). The first volume of *Radiocarbon* (the *Radiocarbon Supplement*) contained a selected list of publications up to 1958 compiled by Frederick Johnson (1959). Dilette Polach (1980) has initiated the

TABLE 1.1
Major Radiocarbon Conferences to 1987

	A.	International Radiocarbon Conferences		
Conference	Location	Dates	Proceedings	Reference
1st	Copenhagen, Denmark	September 1–4, 1954	none	Goodwin, 1954
	Andover, Massachusetts, USA	October 21–23, 1954	unpublished	F. Johnson, personal communication
2nd	Cambridge, England	July 25–30, 1955	none	Levi, 1955b
3rd	Andover, Massachusetts, USA	October 1–4, 1956	none	Johnson et al., 1957
4th	Groningen, Netherlands	September 14–19, 1959	none	Godwin, 1959; Waterbolk, 1960
5th	Cambridge, England	July 23–28, 1962	none	Godwin, 1962a
6th	Pullman, Washington, USA	June 7–11, 1965	Chatters and Olson 1965	—
	[Monaco][a]	[March 2–10, 1967]	[International Atomic Energy Agency 1967]	—
	[Uppsala, Sweden][a]	[August 11–15, 1969]	[Olsson 1970[a]]	
8th	Wellington, New Zealand	October 18–25, 1972	Rafter and Grant-Taylor 1973	Olsson, 1974b
9th	Los Angeles and San Diego, California, USA	June 20–26, 1976	Berger and Suess 1979	—
10th	Bern, Switzerland and Heidelberg, Germany	August 19–26, 1979	Stuiver and Kra 1980a, 1980b	—
11th	Seattle, Washington, USA	June 20–26, 1982	Stuiver and Kra 1983	—
12th	Trondheim, Norway	June 24–28, 1985	Stuiver and Kra 1986	—

B. Major Symposia on Accelerator Mass Spectrometry

Conference	Location	Dates	Proceedings	Reference
1st	Rochester, New York, USA	April 20–21, 1978	Gove, 1978	—
2nd	Argonne, Illinois, USA	May 11–13, 1981	Kutschera, 1981	—
3rd	Zurich, Switzerland	April 10–13, 1984	Wolfli et al., 1984	—
4th	Niagara-on-the-Lake, Ontario, Canada	April 27–30, 1987	—	—

a Between the Pullman Conference in 1965, which was designated as the *Sixth Conference on Radiocarbon and Tritium Dating*, and the conference held in 1972 in Wellington, New Zealand, identified as the *Eighth International Radiocarbon Dating Conference*— conferences dealing with radiocarbon dating were held in Monaco and Uppsala, Sweden. The Monaco conference, however, was considered by its organizers to be a successor to an International Atomic Energy Agency (IAEA) conference held in 1962 in Athens. The Uppsala (Twelfth Nobel Symposium) conference was not organized as a general radiocarbon conference (I. U. Olsson, personal communication). However, the organizers of the Wellington conference looked back to the Uppsala conference as the seventh in the general series of radiocarbon conferences (Rafter and Grant-Taylor, 1973:iii).

compilation of a bibliography of radiocarbon dating. She has analyzed the nature of the literature and has determined that the ability to retrieve bibliographic materials dealing with archaeological applications of the ^{14}C method using existing retrieval methods, is, at present, very poor.

Systematic methods of efficient data retrieval for individual ^{14}C determinations have also been slow to develop. Initially, a relatively comprehensive retrieval system for ^{14}C dates was developed by the Radiocarbon Dates Association, Inc. utilizing edge-punched card files employing needle sorting. Unfortunately, the time required to sort through large groups of cards became excessive since the number of individual dates dramatically increased in the 1970s. A very useful index for ^{14}C values associated with archaeological materials was assembled by Jelinek who arranged dates published through 1961 on a geographical basis (Jelinek, 1962). However, neither of these approaches provided for the rapid and accurate retrieval of specific dates from a large group of determinations nor were they designed to cope with the large increase in the number of dates. Several types of ^{14}C data retrieval approaches employing computer-based systems are currently under development (e.g., Taylor *et al.*, 1968; Gulliksen, 1983; Moffett and Webb, 1983; Otlet and Walker, 1983:103).

CHAPTER 2

DEFINITION OF MAJOR ANOMALIES

2.1 GENERAL CONSIDERATIONS

There are two basic components involved in the process of developing or evaluating temporal frameworks based on ^{14}C data. The first is concerned with the contextual parameters that document the relationship of a sample to a specific archaeological or geological feature. Section 5.2 discusses such *sample provenance factors* that focus on the integrity of the association of sample materials with an event or phenomenon for which temporal placement is desired. The second component involves considerations of the degree to which the geophysical and/or geochemical assumptions of the method hold for a particular sample or related set of samples. This component includes three sets of factors that impinge on the accuracy and precision of individual ^{14}C age estimates: *sample composition factors* (variations in carbon isotope ratios due to contamination and fractionation effects), *statistical and experimental factors* (constraints imposed by the nature of radioactive decay and methods of measurement), and *systemic factors* (temporal or secular variations in initial ^{14}C concentrations in all samples as well as reservoir effects due to variability in initial ^{14}C concentrations in different geochemical environments).

Issues relating to sample composition factors are defined later in this chapter and are considered in detail in Chapter 3 and Section 5.3. Statistical and experimental factors, including the basis on which ^{14}C age estimates are inferred from counting and other analytical data, are reviewed in Chapter 4. Of the systemic factors, those dealing with reservoir effects are defined later in this chapter and considered in detail in Section 5.5.1. Most of the remainder of this chapter considers the implications of the existence of secular variations in ^{14}C activity—the major (long term) trend and de Vries (shorter term) effects. At the conclusion of this chapter, the effects of recent additions to the amount of ^{14}C in the exchange reservoirs—the fossil fuel and atomic bomb effects—will be briefly noted.

In critical evaluations of ^{14}C determinations on samples from archaeological contexts, it is important to keep in mind that age estimates based on such data take the form of time intervals or time segments. The magnitude of these time segments is variable, compounded by the various factors noted in the first paragraph. Although the statistical constraints usually contribute the largest—and most easily quantifiable—component of the overall range of uncertainty, sample composition and systemic factors must also be taken into account. For the reasons reviewed in Chapter 5 and particularly in Section 5.6, the minimum overall level of uncertainty for an individual ^{14}C age estimate for middle and late Holocene samples is about 200 years. For early Holocene samples, this minimum value would be about 300 years and at least 500 years for late Pleistocene materials. Only with suites of ^{14}C values involving multiple analysis of duplicate samples or under very unusual circumstances (e.g., the "wiggle matching" approach) can reductions in the minimum overall uncertainties be justified.

2.2 SECULAR VARIATION EFFECTS

One of the most fundamental assumptions of the ^{14}C method is the requirement that natural ^{14}C concentrations in materials of "zero ^{14}C age" in a particular reservoir are equivalent to that which has been characteristic of living organisms in that same reservoir over the entire ^{14}C time scale. Generally this assumption is seen to require an equilibrium or steady-state relationship in which the production and decay rates have been in approximate balance. Since the decay rate of ^{14}C is constant (Section 4.1), the principal variables affecting equilibrium conditions would be changes in the atmospheric production rate of ^{14}C and/or variations in the parameters of the carbon cycle such as reservoir sizes and rates of transfer of ^{14}C between different carbon reservoirs (Damon et al., 1978). To appreciate how equilibrium conditions are achieved, consider a hypothetical situation in which ^{14}C is absent from all carbon reservoirs due to the exclusion of cosmic rays from the vicinity of the earth. If cosmic-ray bombardment at present levels were suddenly initiated, the buildup of ^{14}C could be plotted. This would permit the visualization of the process and time frame by which a balance between decay and production/transfer rates would be achieved.

Figure 2.1 is a representation of this process using a box-diffusion model (Oeschger *et al.*, 1975; Damon *et al.*, 1983; P. Damon and R. Sternberg, personal communication). Following the onset of cosmic-ray bombardment, there would be a relatively rapid rise in the ^{14}C in the atmosphere. However, the rate of increase would be progressively modified as atmospheric ^{14}C was distributed into other carbon reservoirs and the total amount reduced by radioactive decay. As Figure 2.1 illustrates, if the ^{14}C

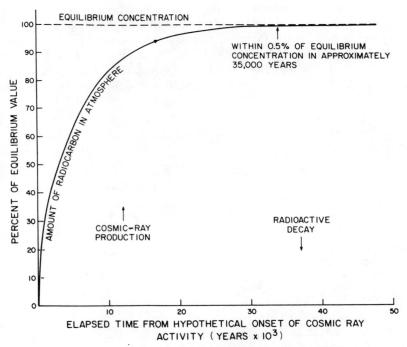

Figure 2.1 Time period for achievement of equilibrium concentration of ^{14}C in the atmosphere employing a box-diffusion model of Oeschger *et al.* (1975) as modified in Damon *et al.* (1983). A general purpose electrical circuit analysis program (SPICE, University of California, Berkeley) was used to generate the ^{14}C concentration values for ten reservoirs (atmosphere, mixed layer in ocean, biosphere, and seven layers in the deep ocean) for various time periods after setting all reservoir ^{14}C concentrations to 0. (Computer analysis performed at the University of Arizona through the courtesy of Paul Damon and Robert Sternberg.)

production and transfer rates remained constant, it would take approximately 35,000 years or about six ^{14}C half-lives for the ^{14}C concentration in the atmosphere to reach within about 0.5% of its equilibrium value. The rate of increase in the biospheric ^{14}C concentration would lag a few percent behind that of the atmosphere for several thousand years. After about 15,000 years, however, the *rate* of ^{14}C increase in the biosphere would be similar to that of the atmosphere. If it can be assumed that modern ^{14}C concentrations in living organisms have been constant over at least the last few hundred thousand years, then organic materials *while living* and deriving their carbon from a common reservoir should exhibit the same ^{14}C activity no matter whether they were alive 100,000, 30,000, or 500 years ago. In such cases, the "zero age" or initial ^{14}C activity for all such samples would be identical and a measurement of their residual ^{14}C content could be directly translated into an actual age.

Obviously, such a procedure would not derive the true age of a sample if, for example, the production or transfer rates had changed in the relatively recent past. Using the same box diffusion model, Fig. 2.2 illustrates the effect of a sudden 10% increase in the ^{14}C production rate. If the new rate and all other parameters remained constant, a new equilibrium in atmospheric ^{14}C activity would be reached at 10% above the "old" equi-

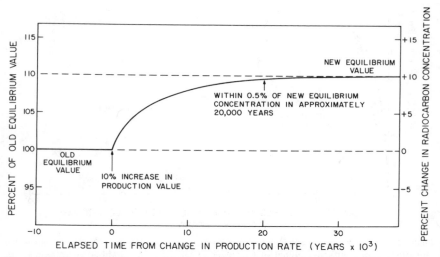

Figure 2.2   Time period for reestablishment of equilibrium condition following 10% increase in production rate using a box-diffusion model of Oeschger *et al.* (1975) as modified in Damon *et al.* (1983). The SPICE program (see Fig. 2.1) was used to calculate the ^{14}C concentration values of the 10% increase for various time periods. (Computer analysis performed at the University of Arizona through the courtesy of Paul Damon and Robert Sternberg.)

librium value in about 20,000 years (P. Damon and R. Sternberg, personal communication). In such a case, a sample drawn from a reservoir that had established this "new" equilibrium would exhibit a ^{14}C-deduced age about 800 years too young with regard to a ^{14}C age calculated on the basis of the old equilibrium value.

The first empirical test of the validity of the assumption of constant concentration and the relatively long-term existence of an equilibrium condition was an analysis of the ^{14}C activity of a series of historically or dendrochronologically documented samples ranging in age from about 1400 to 4600 years (see Chapter 6). The results of these measurements were presented as the first "Curve of Knowns" (Arnold and Libby, 1949). Figure 2.3 is based on the "Curve of Knowns" published in the second edition of Libby's *Radiocarbon Dating* (Libby, 1955:10). The reasonable agreement of the measured values with the theoretical curve strongly supported the initial assumption of constant ^{14}C concentration over at least the recent past. However, because of uncertainties in the values used in the calculations and in experimental conditions in the measurement of ^{14}C in the known-age samples at that time, the intial closeness of fit of the data to the theoretically calculated values was on the order of $\pm 10\%$. In addition, early work involving the comparisons of ^{14}C age values on carbonate materials from deep sea sediment cores with those obtained by the ionium (^{230}Th) method was used to support the assumption that the cosmic ray flux itself had remained essentially constant over much of the ^{14}C time scale (Kulp, 1954b; Broecker and Kulp, 1956).

With the development of ^{14}C counting systems with increased sensitivities, discrepancies between some "known-age" and ^{14}C-derived ages of samples were noted. This stimulated interest in a more systematic investigation of apparent anomalies. Concerns were particularly expressed throughout the 1950s as additional ^{14}C determinations on presumably well-documented Egyptian archaeological samples from the second millennium B.C. were reported. What seemed to be emerging was a general trend suggesting that these values were consistently "too young" by as much as 600–800 years. Interestingly enough, one initial response to these data was a proposal by Libby that the Egyptian chronology for the Old Kingdom Period (ca. 3000–2200 calendar years B.C.) was in error rather than that there was any systematic error in the ^{14}C values themselves (Libby, 1963; Smith, 1964).

In the late 1950s, the Cambridge, Copenhagen, and Heidelberg ^{14}C laboratories obtained ^{14}C determinations on a series of tree-ring samples that showed consistent variation in ^{14}C in the biosphere on the order of several percent (i.e., up to a maximum of about 160 years) over the last 1300 years (Willis *et al.*, 1960). These data confirmed the suggestion of the pioneering

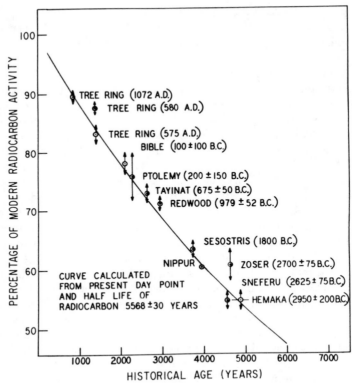

Figure 2.3 Libby Curve of Knowns: Relationship between known age and ^{14}C concentration measured in terms of percentage of modern radiocarbon activity. Taken from Libby (1955:10) but modified by substituting percentage of modern radiocarbon activity for absolute specific radioactivity on the ordinate scale. Previous "Curve of Knowns" had been presented in Libby (1952a:9) and Arnold and Libby (1949:679). Subsequent to the second edition of *Radiocarbon Dating*, the "Curve of Knowns" was included in a number of papers. Beginning with his Nobel Prize lecture paper, however, the ordinate scale was changed to a percentage of modern ^{14}C activity (Libby, 1961b:625). The sample labeled as "Bible" was a portion of the linen wrapping of one of the Dead Sea Scrolls. At the time it was analyzed, the linen was not a "known age" sample since there was a major controversy as to its actual historical age.

Dutch researcher, Hessel de Vries, that "radiocarbon years" and calendar or sidereal years should not be assumed to involve equivalent values (de Vries, 1958). Primarily because of the geophysical implications of these variations, a number of laboratories began to direct their attention to the magnitude and extent of a phenomenon which turned out to have both a major trend (Suess, 1961, 1986) and shorter-term, higher-frequency *(de*

Vries effects) components (Damon and Long, 1962:240; Ralph and Michael, 1974).

Over the last three decades, a number of known-age sample materials have been employed to document the frequency and magnitude of these effects. Although both historically dated materials (e.g., Edwards, 1970; Berger, 1970a; Säve-Söderbergh and Olsson, 1970) and samples from sedimentary deposits (e.g., Stuiver, 1971:58–65) have provided valuable information, the data base that focused attention on ^{14}C secular variation effects was dendrochronologically dated wood, including samples from the giant California sequoia *(Sequoia gigantea),* the European oak *(Quercus* sp.), and particularly the bristlecone pine *(Pinus longaeva)* from Western North America (Pilcher, 1983). The longest continuous tree-ring record has been developed for the bristlecone pine by the late C. Wesley Ferguson of the Laboratory of Tree-Ring Research at the University of Arizona (Ferguson, 1968, 1979; Ferguson and Graybill, 1983).[1]

The oldest stands of *Pinus longaeva* are situated at relatively high altitudes—about 3300 m (10,000 f)—in the White Mountains of east-central California (Fig. 2.4). At this elevation, the period of growth of each annual ring is on the order of 6–8 weeks. Because of this, the tree-ring record is compressed with 1000 years of growth being represented in about 25–30 cm (10–12 in.) of tree radius (Fritts, 1969). The oldest *living* bristlecone pine started growing about 4600 years ago. Because the wood is highly resinous, it is more resistant to physical decay than many other species of wood. Also, at these altitudes, the small amount of ground litter and soil conditions minimize the incidence of fire. The use of remnants of dead bristlecone pine found on the ground surface has extended the period well beyond the life span of any living tree. The tree-ring chronology for the White Mountain bristlecone pine currently extends to 6700 B.C. This represents a continuous tree-ring record of nearly 8700 years. An independently developed tree-ring chronology, also on bristlecone pine but from the upper timberline in the White Mountains, supports the accuracy of the Ferguson chronology at least as far back as about 3500 B.C. (La Marche and Harlan, 1973). Additional "floating" bristlecone tree-ring series containing 500–600 ring segments have been recovered that range in ^{14}C age between 9000 and 11,000 ^{14}C years. This indicates a strong

---

[1] As a matter of historical note, when its great age was first recognized, the bristlecone pine was known as *Pinus aristata* and it is so labeled in early discussions of ^{14}C determinations on bristlecone pine dendrochronologically dated wood. However, in the Great Basin portion of California and Nevada, the tree is now known as *Pinus longaeva*, D. K. Bailey sp. nov. The designation *Pinus aristata* has been retained for the Rocky Mountain form found in Utah, Colorado, Arizona, and New Mexico (Bailey, 1970; C. W. Ferguson, personal communication).

Figure 2.4 View of a bristlecone pine on White Mountains, California. In the background are the Owens Valley and Sierra Nevada range. The increment borer used to obtain samples for dendrochronological examination is shown inserted into the tree. (Photograph courtesy C. W. Ferguson, University of Arizona.)

possibility that a continuous White Mountains bristlecone pine tree-ring chronology can be extended into the Pleistocene (Ferguson and Graybill, 1983; C. W. Ferguson, personal communication). A search for bristlecone pine wood remnants to extend and strengthen the existing series is being actively pursued (Michael, 1984). A relatively low-altitude (<400 m) western European Irish oak tree-ring series, which provides a continuous record spanning almost 7300 years, permits a comparison with the bristlecone pine record (Pilcher et al., 1984).

Radiocarbon determinations primarily on 10-year (decadal) samples (to average out possible problems associated with 11-year solar cycles) of bristlecone pine wood (occasionally 5- or 20-year segments were used) as well as on wood taken from the California giant sequoia and European oak (Becker, 1983; Baillie et al., 1983; Linick et al., 1985) have been undertaken by several laboratories. The data produced as part of these projects have yielded a number of charts, correction tables, and data plots, all designed to provide archaeologists and others with the tools to "calibrate" individual radiocarbon dates. Unfortunately, not all of these plots and corrections agree in detail and some questions have been raised concerning the intercomparability of calibration data from different laboratories. Recent comparisons particularly for the time scale of the last 2000 years obtained by several high-precision laboratories (i.e., those publishing one sigma counting precisions for this period of less than ±20 ^{14}C years) have shown very good agreements (Stuiver, 1982; Pearson and Baillie, 1983). It should be noted that concerns about the possibility that the ^{14}C record in a tree-ring would be compromised by carbon-bearing materials moving between rings appears not to be a major problem when appropriate pretreatment of the wood samples is carried out (Fairhall and Young, 1970; Berger, 1970a, 1973; cf. McPhail et al., 1983).

Five illustrations of calibration data for the ^{14}C time scale are presented in Figs. 2.5–2.9. Figure 2.5 is based on data obtained at the University of California, San Diego (La Jolla/Mt. Soladad), ^{14}C laboratory (Suess, 1970, 1979). Figure 2.6 presents data assembled by the University of Pennsylvania Museum Applied Science Center for Archaeology [MASCA] (Ralph et al., 1973). Figure 2.7 represents ^{14}C determinations on dendrochronological samples or samples associated with historic/calendric data as presented by Clark (1975). Figure 2.8 represents data prepared as a result of a workshop on the calibration of the ^{14}C time scale conducted at the University of Arizona in 1978 (Klein et al., 1982). Finally, Fig. 2.9 presents data obtained on the Irish oak dendrochronological series (Pearson et al., 1986). In addition to these representations of the calibration data, Stuiver (1982, 1983) has derived a high-precision calibration curve for the last 2000 years based on dendrochronologically dated

CONVENTIONAL RADIOCARBON DATES IN RADIOCARBON YEARS BEFORE PRESENT

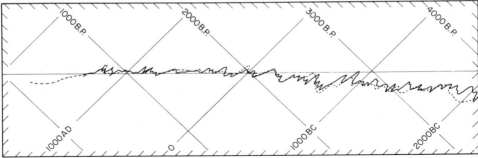

BRISTLECONE PINE DATES IN CALENDAR YEARS

CONVENTIONAL RADIOCARBON DATES IN RADIOCARBON YEARS BEFORE PRESENT

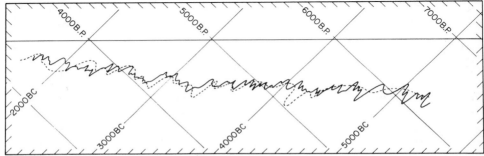

BRISTLECONE PINE DATES IN CALENDAR YEARS

Figure 2.5 Secular variation/major trend (Suess): Relationship of ^{14}C and dendrochronological age of bristlecone pines. Format modified from original source. ^{14}C half-life used = 5568 years; point of origin of plot, A.D. 1950. [Dotted line from Suess (1970), solid line from Suess (1979).]

wood samples—Douglas fir *(Pseudotsuga menziesii)* and California sequoia—from western North America.

2.2.1 Major Trend

In any plot of the ^{14}C/dendrochronological data, the most obvious feature is the long-term or major trend in the secular variation data. It has been suggested that this phenomenon over about the last 8000 years can be represented as a sine wave with a maximum deviation of about 8–10% (650–800 years) at about 4500–5000 calendar years B.C. This apparent sinusoidal feature is illustrated in Fig. 2.6 by a dashed line and in Fig. 2.8 by a solid line. Reviewing the data as presented in any of these plots

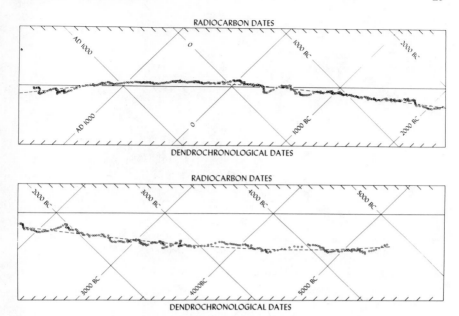

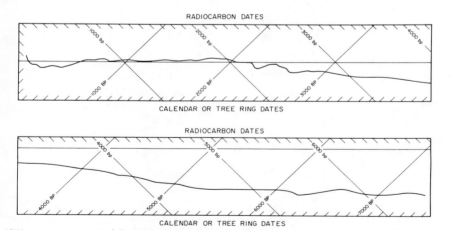

Figure 2.6 Secular variation/major trend (MASCA): Relationship of ^{14}C and dendrochronological age of samples. ^{14}C half-life used = 5730 years; point of origin of plot, A.D. 1950. [After Ralph *et al.* (1973).]

Figure 2.7 Secular variation/major trend (Clark): Relationship of ^{14}C and calendar/dendrochronological ages. Format modified from original source. ^{14}C half-life used is 5568 years; bp indicates ^{14}C values, BP indicates calendar or tree-ring dates. [After Clark (1975), figure 1.]

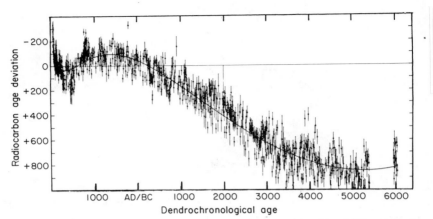

Figure 2.8 Secular variation/major trend (Workshop): Relationship of ^{14}C and dendro-chronological age of wood samples. [After figure 1 in Klein *et al.* (1982).]

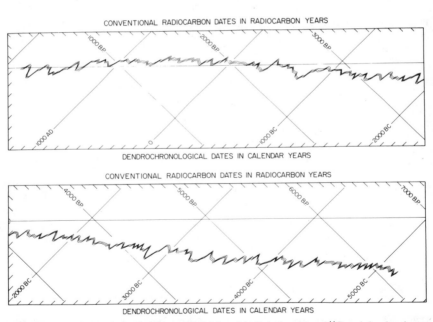

Figure 2.9 Secular variation/major trend (Irish oak): Relationship of ^{14}C and dendrochron-ological age of Irish oak. Format modified from original source. ^{14}C half-life used = 5568 years; point of origin of plot A.D. 1950; beginning of calibration values A.D. 1840. [After Pearson *et al.* (1986).]

clearly illustrates the fundamental fact that radiocarbon years and calendar years are not necessarily equivalent. If such had been the case, all of the data points plotted in these figures would tightly cluster along the central horizontal lines. However, some of the points lie above the lines, indicating in these plots of the data that the ^{14}C values for these periods are "too old" whereas those lying below the lines are "too young." As illustrated in Fig. 2.8, the amount of correction required to bring middle and late Holocene ^{14}C values into approximate alignment with calendar time, as documented by the dendrochronological data, varies from a maximum of about -250 years in the middle of the first millennium A.D. to an apparent maximum of about $+800$ years in the fifth millennium B.C.

Although a number of causes for the observed variations in natural ^{14}C concentrations have been proposed, they can be grouped into two major types as indicated in Table 2.1. The first involves changes in the production rate of ^{14}C due to variations in the intensity and/or composition of galactic cosmic rays, changes in the magnetic field or other characteristics of the sun (heliomagnetic effects), or variations in the characteristics of the earth's magnetic field (geomagnetic effects). The second involves world-wide variations in parameters of the carbon cycle (e.g., reservoir sizes, exchange rates between different reservoirs, amount of CO_2 in the atmosphere) caused by environmental or climatic factors (Siegenthaler *et al.*, 1980). Until recently, the general view was that the principal cause for most of the long-term component or major trend in the secular variation

TABLE 2.1
Suggested Causes of Secular Variation Effects[a]

I. Variations in rate of ^{14}C production in the atmosphere
   1. Cosmic-ray flux variations through the solar system.
   2. Cosmic-ray flux modulation by solar activity.
   3. Cosmic-ray flux modulation by earth's magnetic field.
II. Variations in ^{14}C exchange rates between carbon reservoirs and relative and total CO_2 content of reservoirs
   1. Temperature variations changing CO_2 solubility, dissolution rates, and residence times.
   2. Sea-level variations changing ocean circulation and CO_2 capacity.
   3. CO_2 assimilation variations in terrestrial biosphere in response to changes in biomass and CO_2 concentration.
   4. CO_2 assimilation variations in marine biosphere in response to changes in ocean temperature, salinity, nutrients, up-welling of CO_2-rich deep water, and turbidity of mixed layer of the ocean.
   5. CO_2 injection rate variations into atmosphere from volcanism and other processes.
   6. Sedimentation and biological productivity rate variations in the oceans.

---

[a]Sources: Grey and Damon, 1970; Damon *et al.*, 1978; Libby, L. M., 1973; M. Stuiver, personal communication.

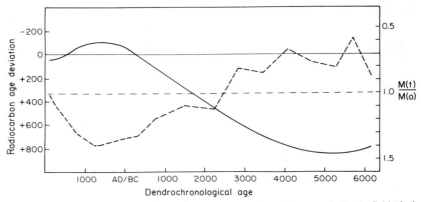

Figure 2.10 Suggested relationship between intensity of earth's magnetic dipole field (dashed lines) and secular variation major trend (solid line). Intensity of field expressed in terms of the means of consecutive 500-yr intervals of the virtual axial dipole moments (VADM) $M(t)$, relative to the modern value $M(0)$ for the last 7000 ^{14}C yrs. [Data from figure 2(b) in Barton *et al.* (1979). Method of representing VADM values from figure 3 in Stuiver and Quay (1980). Secular variation curve, in inverted form, from figure 1 in Klein *et al.* (1982); see figure 2.8. Cf. Bucha (1970: 508).]

effects had been changes over time in the intensity of the earth's geomagnetic dipole field.

Increasing the strength of the dipole field would result in increased shielding of the earth's atmosphere from cosmic radiation and a corresponding reduction in ^{14}C production. Radiocarbon concentrations in materials growing during such periods would be correspondingly reduced and this reduction in ^{14}C activity would translate into inflated ^{14}C ages. Conversely, a decreased dipole field would result in samples exhibiting anomalously young ^{14}C age values. Figure 2.10 presents suggested changes in the Holocene dipole field strength (Barton *et al.*, 1979; Stuiver and Quay, 1980) and the main trend of changes in ^{14}C activity in the biosphere over the last 7000 years as represented in Fig. 2.6 (Klein *et al.*, 1982). Several investigators have previously cautioned that much additional study is required to more fully document the apparent sinusoidal nature of the dipole field over the last 7000 years. Before about 7000 years B.P. the nature of the geomagnetic intensity data is even more problematic. Any extrapolation of the ^{14}C secular variation curve based on the inferred cyclic pattern of the dipole field in the Pleistocene (e.g., Bucha, 1967) is especially fraught with difficulties because of uncertainties in several geophysical and geochemical parameters. Potentially the most severe anomaly would result from the effects of geologically recent reversals or major excursions in the earth's magnetic field.

During a reversal of the earth's dipole field, the intensity would be significantly reduced before the field built back up to the opposite polarity mode. An excursion in the dipole field involves variations in intensity and direction that do not result in a change in polarity. It has been suggested that the length of time it takes for a full polarity shift or major excursion to occur seems to be on the order of several thousand years. However, the exact mechanisms that would account for time scale for the reversals are the subject of current debate. One recent view offered by R. A. Muller and D. E. Morris (personal communication) suggests that they are driven by extraterrestrial impact events or volcanic eruptions.

Obviously, there would be a significant effect on ^{14}C production during a period of reversal or major excursion. Several investigators have suggested the presence of at least one major excursion in the dipole field within the last 50,000 years (e.g., Barbetti and McElhinney, 1972; Morner, 1977). However, it is uncertain whether such a phenomenon actually involves the dipole field or might rather represent regional or continental anomalies affecting only a portion of the earth's surface. It is necessary to await further clarification in the interpretation of the data (Wolfman, 1984:372–374; Aitken, 1985:213–215). Such uncertainty in this and other geophysical factors has made it necessary to proceed with great caution in "calibrating" ^{14}C determinations in the absence of clear known-age controls for the period prior to that which can be documented by dendrochronological/^{14}C data.

Questions concerning the relationship between variations in the dipole field and the main trend secular variation effect have recently been raised as a result of measurements of the concentration of a radioactive isotope of Beryllium, ^{10}Be. Most ^{10}Be is produced by cosmic ray reactions on nitrogen and oxygen and has a half-life of about 1.5 million years. Few measurements of natural concentrations of ^{10}Be had been undertaken until the development of AMS counting techniques (Section 4.5). It is deposited on the surface of the earth by precipitation and is preserved in snow and ice layers, in soils, and in ocean/lake sediments (Beer *et al.*, 1983). If there had been long-term secular changes in geomagnetic intensity over the period of the Holocene, this effect should be reflected in changes in ^{10}Be concentration over time paralleling the changes observed with ^{14}C. Initial interpretations of ^{10}Be measurements in ice cores from Greenland spanning the last 4000 years seemed to show no evidence of a major trend (Beer *et al.*, 1984). A reinterpretation of this data by other workers (Damon and Linick, 1986) argues that secular variation effects can indeed be seen in the ^{10}Be measurements. Further work to clarify these data is continuing.

Other approaches to ascertaining the nature and magnitude of secular variation effects for the period before dendrochronological calibration data are available and include attempts to correlate ^{14}C age estimates with those based on lake sediments/varves, thermoluminescence (TL), and uranium

(U)-series determinations and the inferred cyclical pattern in the nondipole component of the geomagnetic field. Varve data from Sweden were used to support the suggestion that major secular variation deviations decrease as the Holocene/Pleistocene boundary is approached (Tauber, 1970). However, a cautionary note was later sounded that the temporal scale for the Swedish varve chronology may be more uncertain than had been originally thought (Tauber, 1980). Radiocarbon measurements on lake sediments from North America (Lake of the Clouds) did not support the inferences based on the Swedish varve data (Stuiver, 1970). Thermoluminescence data, as assembled by Barbetti (1980), suggest that ^{14}C values ranging around 30,000 ^{14}C years are registering up to several thousand years too young. However, the relatively large errors associated with these TL values render the comparisons difficult to evaluate. U-series age estimates obtained by Vogel (1983) seem to indicate that ^{14}C ages may register several thousand years too young at about 18,000 and 30,000–40,000 years ago. Data assembled by Stuiver (1978b), which included summaries of nondipole geomagnetic profile studies from sediments in Lake Windermere and the Black Sea regions, suggested that ^{14}C age anomalies in the range of 9000 to 30,000 ^{14}C years appear not to exceed 2000 years. Despite such encouraging data, the interpretation of ^{14}C values *in absolute terms* from these time periods needs to be approached with due regard to the nature of the uncertainties. Such variations would not, however, affect the geographical and temporal intercomparability of ^{14}C values and the ability to use ^{14}C data from the Pleistocene to provide accurate ordering on a serial time scale.

2.2.2 De Vries Effects

If we examine the plots in Figs. 2.5–2.9, we note that each exhibits apparent shorter-term, higher-frequency variations superimposed on a longer-term trend. Historically, short-term perturbations were initially noted by de Vries (1958) and early studies had distinguished long- from shorter-term variations (e.g., Suess, 1965; Stuiver and Suess, 1966). However, the fact that there appeared to be an irregular series of "wiggles" (originally "wriggles"), "kinks," or "windings" in the calibration curve—some of them of major amplitude—was first widely discussed as a result of data presented by Suess (1970:309–310; plate 1).[2] Since Suess (1970:310) noted that his original "wriggles" had been drawn by a subjective process, which he called "cosmic schwung," a reluctance was expressed by some

---

[2]The terms "Suess wiggles" and "medium-term variations" have been used to refer to most of what is here labeled as "de Vries effects" (de Jong *et al.*, 1979; Mook, 1983b).

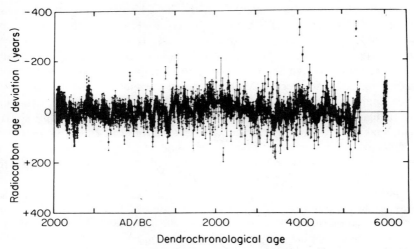

Figure 2.11 Secular variation/de Vries effect: Short term variation in ^{14}C activity. Plot results from the subtraction of major trend function. Based on composite plot of data. [After Klein *et al.* (1980).]

to accept the magnitude of a number of the proposed shorter-term deviations. For example, it was suggested that some of the major anomalies presented by Suess were a product of statistical variance in the ^{14}C data (e.g., Clark and Renfrew, 1972:7; Pearson *et al.*, 1977:28). There is now a consensus that de Vries effects reflect geophysical reality and even that Suess's proposed large deviations in several time intervals (e.g., Maunder minimum) are of the correct magnitude (e.g., de Jong *et al.*, 1979; Suess, 1980; Pilcher, 1980; Stuiver, 1982; cf. Berger, 1985). Figure 2.11 presents one plot of de Vries anomalies over the last 7000 years based on the composite bristlecone pine ^{14}C data. This plot results from the subtraction of the main trend of an apparent sine wave function from each data point. These residual values can be seen to vary around the main trend line as defined in Klein *et al.* (1980).

Factors which have been most often suggested as causes of the de Vries variations include solar or heliomagnetic modulations of ^{14}C production rates (Stuiver and Quay, 1980) and the effects of a nondipole component on the geomagnetic field (Thompson, 1983:229; Damon and Linick, 1986). The suggested heliomagnetic process involves changes in the solar corona that affect the solar wind, which in turn deflects cosmic rays in the vicinity of the earth. The most obvious physical feature associated with variations in solar activity are changes in the number of sunspots over time. Several sunspot cycles have been identified, the best known being the 11-year episode. A significant correlation has been found between the cosmic-ray flux as observed in the upper atmosphere and the 11-year cycle. Observations made over the last few centuries have recorded sunspot numbers and these records have been compared to variations observed in the bristlecone pine tree-ring record. Although good correlations can be obtained, it has been argued that a much better fit of the data has been achieved by correlating variations in ^{14}C activity with an index reflecting short-term, magnetic variations on the sun (Castagnoli and Lal, 1980). It has been suggested that what, at first glance, appears to be a series of *irregular* short-term ^{14}C fluctuations is actually revealed, upon further analysis, to contain distinct periodicities (Suess, 1986).

The existence of both long-term (major trend) and shorter-term (de Vries) secular variation effects bears directly on the accuracy and precision of ^{14}C determinations. We can, for example, determine from any of the four plots of data that samples exhibiting an age value of, for example, 4700 ^{14}C years are actually a little less than 800 years too young as measured on the dendrochronological time scale. This might be interpreted to mean that to convert ^{14}C years to calendar years for these samples, one would simply add the difference so as to provide a calibrated ^{14}C value. Unfortunately, this approach would ignore two important interrelated factors. First, all ^{14}C age expressions, including any ^{14}C value to be calibrated and the calibration data itself, inherently carry with them a statistical uncertainty. A ^{14}C age expression is a *time interval* within which there is a given probability that the actual age of the sample lies. (The nature of statistical error in the evaluation of ^{14}C values is discussed in Sections 4.7 and 5.4). Second, a simple additive approach would ignore the effect of shorter-term variations in ^{14}C values. Fig. 2.12 presents the relationship between ^{14}C and dendrochronological age values for the interval of 4500 to 5100 ^{14}C years B.P.

If we examine Fig. 2.12, we note that in several instances (e.g., at about 4500 and especially at about 4700 ^{14}C years B.P.), a ^{14}C value can have multiple dendrochronological (= calendric) equivalents. This means that, for some periods, "radiocarbon time" is elastic, i.e., it can be stretched out or foreshortened. It is important to observe that the de Vries effects documented in Fig. 2.12 have been constructed from ^{14}C values with very small statistical errors, on the order of ±15–30 years. Such precision usually requires relatively large sample sizes (7–20 g of carbon). Such conditions can generally not be met with the typical archaeological sample. Statistical variances of between ±50 and ±100 years for Holocene samples (less than 10,000 years) are more common. If we assign a one sigma statistical error of ±100 years to the value of 4700 ^{14}C years, we see that the equivalent dendrochronological age at that level of confidence

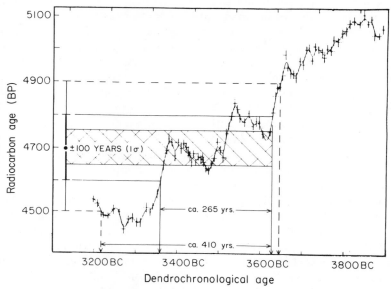

Figure 2.12 Secular variation/de Vries effect: Short term variations in ¹⁴C activity for the period from approximately 4500 to 5100 ¹⁴C years B.P. Dendrochronological time scale based on European oak (*Quercus* sp.) from Southern Germany. ¹⁴C half-life used is 5568 years. [Dendrochronological and ¹⁴C data plot from figure 1 in de Jong and Mook (1980).]

ranges over about a 265-year period between about 3360 and 3625 calendar years B.C. If we double the statistical error to increase the confidence level (from 68% to 95%), the equivalent calendric age ranges over a period of about 410 years. Interestingly enough, if we reduce the statistical error to ± 50 years (the crosshatched area), the resulting calibration range is not significantly reduced below that with ±100 years, due to a more rapid change in ¹⁴C activity about this time.

2.3 RESERVOIR EFFECTS

One of the most important characteristics of the ¹⁴C method is its ability to provide comparable age estimates for organic materials on a *worldwide* basis. For the ability to be realized, ¹⁴C must be mixed rapidly (within a few years) and completely throughout all of the carbon-containing reservoirs. If such a condition prevailed, the contemporary ¹⁴C content of all organic samples would be essentially identical. Early in the history of ^{14}C dating it was determined that, in a number of instances, this was not the case. One of the earliest illustrations of the breakdown of this assumption was the determination that *living* samples from a fresh water lake with a limestone bed exhibited apparent ^{14}C "ages" of as much as about 1600 years. In this situation, the ^{14}C content of the source from which the materials drew their carbon reflected the atmospheric ^{14}C activity as modified by the lack of ^{14}C activity in $CaCO_3$ (limestone), which in this case was Paleozoic in origin (with an age in excess of several hundred million years) and thus contained no measurable ^{14}C. Thus, the "contemporary" ^{14}C activity of carbonates and biocarbonates in this particular lake environment was about 20% below the contemporary level in the ordinary biosphere (Deevey *et al.*, 1954).

The overwhelming bulk of sample materials encountered in archaeological applications derives from plant and animal material from the terrestrial biosphere and organic (fish, marine mammals) and inorganic (marine shell carbonates) materials from near-shore ocean environments. Some of the problems that are associated with the interpretation of ^{14}C values derive from variations in initial ^{14}C concentrations in different carbon reservoirs such as those found in terrestrial as opposed to marine environments.

As we have noted, most ^{14}C laboratories use as their modern standard either preparations of oxalic acid distributed by the U.S. National Bureau of Standards or a secondary standard whose ^{14}C activity in relation to the NBS oxalic acid standards is known (Section 4.6). The modern or contemporary ^{14}C activity of terrestrial organics is inferred in terms of these standards. Anomalies result when the initial ^{14}C activity in a given sample is not equal to that contained in standard terrestrial organics. Problems involving these reservoir effects are discussed in Section 5.5.1.

2.4   CONTAMINATION AND FRACTIONATION EFFECTS

For the ^{14}C method, the basic physical phenomena used to index time are changes in $^{14}C/^{12}C$ ratios. If carbon had only two naturally occurring isotopes, only these two isotopes could enter into calculations. However, as we have already noted, this element has *three* naturally occurring isotopes existing in modern organics in the ratio of approximately 99% ^{12}C, 1% ^{13}C, and 10^{-10}% ^{14}C; i.e., for every ^{14}C atom there are about 10^{10} ^{13}C atoms and 10^{12} ^{12}C atoms. Accurate estimates of the age of a sample using the ^{14}C method assume that no change has occurred in the natural carbon isotopic ratios except by decay of ^{14}C. Several physical effects other than radioactive decay, however, can alter the carbon isotope ratios in samples. The most commonly discussed problem involves *contamination* of samples in which carbon-containing compounds not indigenous to the original organic material are physically or chemically incorporated into a sample matrix. A second problem involves the *fractionation* of the carbon isotopes under natural conditions. Fractionation involves alterations in the ratios of isotopic species as a function of their atomic mass. In this context, ^{14}C has a mass about 15% greater than ^{12}C, i.e., ^{14}C is "heavier" than ^{13}C or ^{12}C. During certain natural biochemical processes (e.g., photosynthesis) "lighter" isotopes are preferentially incorporated into sample materials. Because of this, variations in ^{14}C/^{12}C ratios can occur that have nothing to do with the passage of time. Chapters 3 and 5 consider these topics. Isotopic fractionation can also occur in the laboratory during the processing of samples (Chapter 4).

2.5 RECENT ^{14}C VARIATIONS

Several factors contribute to making it very difficult to employ ^{14}C data to assign unambiguous calendar age estimates to materials dating to the last 300 years. Unfortunately, de Vries effects are particularly pronounced for this period. In addition, during the nineteenth and twentieth centuries, ^{14}C concentrations were seriously affected by human activities, first, as the result of the combustion of fossil fuels (*Suess* or *industrial effect*) and then, in the post-World War II period, as a result of the detonation of thermonuclear devices in the atmosphere (*atomic bomb* or *nuclear effect*). Because of the combination of these effects, a ^{14}C age estimate may be reported as simply *modern* when the reservoir-corrected, conventional ^{14}C value is less than 200 years (Stuiver and Polach, 1977:362).

2.5.1 Post-Sixteenth-Century de Vries Effects

Figure 2.13 illustrates de Vries effects for the last 400 years of ^{14}C time using the high-precision data of Stuiver (1982). Beginning about A.D. 1650, significant natural ^{14}C variations created a situation in which it is not possible to assign an actual calendar age to any sample derived from this time period to better than about a 300-year time span unless "wiggle-matching" procedures are employed (Section 5.6). A ^{14}C value of 150 ± 40 years with a one and two sigma range has been indicated in Fig. 2.13.

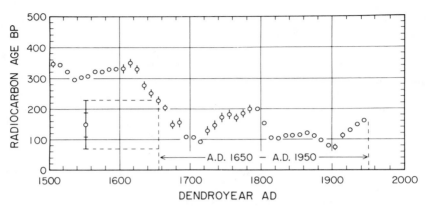

Figure 2.13 Recent de Vries effects: A.D. 1500 to A.D. 1950. Dendrochronological scale based on Douglas fir (*Pseudotsuga menziesii*) from the Pacific northwestern portion of the United States. [From figure 2A in Stuiver (1982).]

Clearly, the only age expression that could be inferred for this sample would range over the whole 300-year span of time. In this case, significantly decreasing the statistical variance for this sample would result in a very marginal increase in the precision of the inferred calendar age.

2.5.2 Suess Effect

Early workers in the radiocarbon field found that ^{14}C activity in late nineteenth-century wood and certain other terrestrial organics was depleted with respect to the expected ^{14}C activity by up to several percent. The primary source of the problem was traced to the combustion of fossil fuels particularly in the twentieth century. Because of their great geological age, fossil fuels such as coal, oil, and natural gas contain no measurable amounts of ^{14}C and are referred to as "dead" in terms of their ^{14}C activity. Fossil fuel combustion dilutes ^{14}C concentrations since it adds CO_2 to the atmosphere that contains no ^{14}C. The approximate magnitude of this dilution was first documented systematically in the early 1950s by Hans Suess, then of the United States Geological Survey (Suess, 1955). A small portion of the deviation observed in ^{14}C activity during this period is superimposed on the natural ^{14}C variations (de Vries effects). Figure 2.14 illustrates the Suess effect for the period from 1900 to 1955 (Cain and Suess, 1976; Levin, *et al.*, 1985). One result of the identification of the Suess effect was the recognition that materials growing during this period could not be used to define the natural "0 B.P." ^{14}C activity without correcting for recent artificial and natural changes in ^{14}C activity in modern

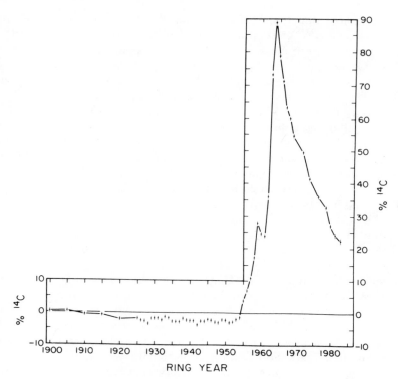

Figure 2.14 Suess and atomic bomb effects: ^{14}C activity in tree rings (1900–1955) and tropospheric CO_2 (1955–1983) expressed as percentage above or below 0.95 NBS oxalic acid standard (old). Tree ring ^{14}C data (solid dots with error bars) obtained from samples collected at Bear Mountain State Park, New York (41°N, 74°W), reported in Cain and Suess (1976:3691). Tropospheric ^{14}C data (solid squares) from CO_2 collected at Vermunt, Austria (47°N, 10°E), reported in Levin *et al.* (1985:18) represent yearly means. ^{14}C values are expressed as percentage above or below contemporary standard rather than in per mil as in original citations.

biologicals. This was one factor that led to the introduction of contemporary standards such as NBS oxalic acid.

2.5.3 Atomic Bomb Effect

Since the detonation of nuclear and thermonuclear devices yields large thermal neutron fluxes, the testing of these weapons in the atmosphere has produced significant amounts of artificial or "bomb" ^{14}C (Rafter and Fergusson, 1957). Between 1955 and 1963, the ^{14}C activity in terrestrial organics almost doubled. Unfortunately, it was during this period that early work on the ^{14}C method, using solid or particulate carbon, occurred. Fallout of fission products from bomb testing complicated the problem of doing accurate low-level ^{14}C counting. An international agreement in 1963 halted testing in the atmosphere by most of the nuclear powers. This allowed ^{14}C to begin the process of reestablishing a new atmospheric ^{14}C equilibrium. If this had been the only source of artificial ^{14}C, then the new equilibrium condition would have been about 3% above pre-1950 levels assuming that no further atmospheric testing occurs (Fairhall and Young, 1985). However, by this time, the continuing combustion of fossil fuels will probably have compensated for this increase (Stuiver and Quay, 1981; M. Stuiver, personal communication).

The rapid injection of artificial ^{14}C into the carbon cycle allowed it to be used as a "tracer" to determine contemporary mixing mechanisms and rates of exchange ("turn-over-times" or "residence times") between different carbon reservoirs (Broecker and Peng, 1982). Since much of the bomb radioactivity was added at high altitudes, there was a similarity between natural and artificial ^{14}C injection points. It was determined that mixing between the stratosphere and troposphere takes up to about 2–4 years. Residence times in the atmosphere, involving exchange with the surface ocean, range between 6 and 10 years whereas exchange between surface and the deep ocean is much longer, in some cases in excess of several hundred years. Some bottom water contains inorganic carbonates with ^{14}C ages of about 1200 years [M. Stuiver, personal communication; for early literature see Fairhall and Young (1970)]. These data confirmed the belief that mixing rates for ^{14}C in most terrestrial carbon reservoirs is sufficiently rapid so that the maximum variability in "contemporary" activity for terrestrial carbon samples is on the order of ±40 years. This is also the lower end of routine precision for most standard ^{14}C determinations on materials less than about 5000 years old.

CHAPTER 3

SAMPLES AND SAMPLE PRETREATMENT

3.1 GENERAL PERSPECTIVE

The accuracy of a ^{14}C determination depends on both the degree to which the association between sample material and the event or phenomenon to be dated is direct and unambiguous and the degree to which the geophysical assumptions of the method hold for a particular sample. Illustrations of the problems of association or context are discussed in Chapter 5. This chapter reviews issues related to the use of various types of organics in terms of physical and chemical characteristics that affect their use as samples for ^{14}C analysis. In evaluating this aspect of the integrity of sample materials, it is necessary to consider the possibility of both *contamination* effects and *fractionation* effects.

Contamination is a process involving the incorporation of nonindigenous organics into a sample matrix. The effects on ^{14}C values resulting from the introduction of foreign carbon depends on the nature and condition of the sample material, the characteristic(s) of the geochemical environment(s) within which the sample has been embedded, and the time frame over which such action(s) occurred. Each situation may be unique, and precautions exercised to avoid contamination effects may have to be designed to fit each particular situation. While recognizing that exceptions to general characterizations occur, a few generalizations and accepted routine procedures in the pretreatment of samples have been established by most laboratories. These procedures are sample-type specific and are generally concerned with removing what is assumed to have not been present when the original sample was removed from its carbon reservoir.

It is usually possible to infer the effect of known contamination effects on a given sample in terms of the direction the age change will take (i.e., making the apparent age too old or too young), but the specific magnitude of an anomaly depends on the true age of the original sample, the age of the contaminant, and the percentage contribution of the contaminant. Rarely can all of these factors be known. However, the effect of varying percentages of contamination of known activity on samples of differing age can be calculated to obtain an estimate of the order of magnitude of potential error that might be introduced. Observations made by archaeologists while excavation is in progress can greatly assist laboratory personnel in the application of appropriate sample pretreatment procedures. Examples of relevant observations include the reporting of widespread rootlets in the vicinity of the sample collection zone, the presence of high organic content materials such as peats, and the proximity of petroleum products such as asphalt or tar or fossil organics such as lignite or coal. Details of how evaluations of contamination factors may be carried out are discussed in Chapter 5.

Fractionation can result from natural variations in carbon isotope values. It can also occur during laboratory preparation of samples if certain precautions are not observed. It has been determined that variations in stable carbon ratios ($^{13}C/^{12}C$) have a predictable effect on the $^{14}C/^{12}C$ ratio (Craig, 1953). This fact violated one of the fundamental assumptions on which the ^{14}C method rests, namely, that there were no effects that would alter the $^{14}C/^{12}C$ ratio other than that caused by the decay of ^{14}C. Subsequent $^{13}C/^{12}C$ measurements (typically expressed as $\delta^{13}C$) have determined that a considerable range in $\delta^{13}C$ values is exhibited by organic materials. Terrestrial and marine organisms particularly exhibit a significant divergence in their $\delta^{13}C$ values. A general consensus has developed regarding a set of standards and procedures to normalize samples with different $\delta^{13}C$ values onto a common scale. Details of how these methods are applied are reviewed in Section 5.3.2. Fractionation can be avoided in sample processing by ensuring that all processes are carried out with very high (>95%) yields. If all of a sample is converted from one form to another (i.e., from a solid to a gas), the isotope ratios will remain unchanged.

3.2 SAMPLE PRETREATMENT STRATEGIES

The purpose of sample pretreatment is to remove any carbon-containing compound that might exhibit ^{14}C activities different from that of the indigenous sample material. Major contaminants can be grouped into four broad categories (i) *physically removable,* (ii) *acid soluble,* (iii) *base soluble,* and (iv) *solvent soluble.* In each category a number of techniques have been developed to facilitate removal of nonindigenous carbon from samples. Procedures involved in the conversion of a pretreated sample into a form that conforms to the requirements of a particular counting system are reviewed in Chapter 4.

The first step in any laboratory pretreatment is a careful visual examination of the sample and removal by physical means of obvious surface and/or intrusive nonsample material, e.g., rootlets and micro-organisms. With relatively solid macrosample materials, it may be advantageous to remove the surface layer by cutting. Close visual examination is important in documenting the physical character of the sample material. A primary initial concern should be to precisely identify the nature of the carbon-containing compounds of which the sample is composed. Almost all sample types can be taxonomically or biochemically identified by appropriate techniques. When possible, the proper scientific nomenclature for species of plant and animal sample material should be obtained even if the fragmentary nature of the sample permits only genus or even family level designations. In areas where sources of fossil carbon were readily available in surface exposures or relatively shallow deposits and were known to have been employed by aboriginal populations (e.g., tar, bitumen or asphalt to coat basketry or the use of coal as a source of fuel), it is obviously critical that such materials be identified and removed from sample preparations (cf. Evin, 1983). Samples from such areas that are composed of small bits of carbonized organics and labeled by the collector as "charcoal fragments" should be particularly scrutinized.

Terrestrial organic samples, such as charcoal and wood, deposited in a soil environment, even for a relatively short period of time, generally will have absorbed varying amounts of inorganic carbonates, particularly calcium carbonate ($CaCO_3$), from percolating groundwater. The use of acidic solutions, generally hydrochloric acid (HCl), to remove absorbed carbonates is the most widespread pretreatment employed. Such a pretreatment is necessary because the dissolved $CaCO_3$ generally exhibits ^{14}C activities that are significantly depressed from contemporary levels. For both marine and freshwater gastropods, acid etching of the surface is typically employed to remove the surface layers where isotopic exchange is likely to be present.

Active soil horizons in most temperate regions of the world contain decay products of organic substances sometimes labeled "humic acids" and "fulvic acids." The sources and precise nature of these humic substances are complex and, to a degree, still incompletely characterized (Ertel *et al.*, 1984). These compounds are distributed through soil profiles by ground water action and can be absorbed into sample materials. The fulvic acid component is acid soluble, whereas humic acids are acid-insoluble. However, it can be solubilized by treatment with base solutions such as sodium hydroxide (NaOH). It should be noted that extractions with NaOH or other base solutions must be carefully carried out due to the possibility of exchange with atmospheric CO_2 while a sample is undergoing pretreatment. Typically such a problem is avoided by reacidifying the sample after the base solution is removed by filtering and washing. In some cases, a filtrate from the NaOH solution can be precipitated with acid and this fraction examined separately to determine if the "humic acid" component exhibits a significantly different ^{14}C activity.

In special situations, potentially important sample materials may have been impregnated with organic materials that can only be removed with appropriate solvents such as ether, benzene, methanol, or toluene. Examples of such necessary pretreatment include samples derived from petroleum deposits. Although the effects of contamination with petroleum products on the ^{14}C age of relatively young materials can be significant, such situations fortunately are rare. An extreme case has been reported on charcoal samples from an archaeological site at Terqa, Syria, recovered from a level dated by two charcoal samples and by the archaeological context to about 3000 B.C. The age of a third charcoal sample recovered from the same context was $28,700 \pm 1100$ ^{14}C years B.P. (Venkatesan *et al.*, 1982). Extensive extractions with organic solvents and chemical analysis strongly point to extreme contamination with asphalt, which was used in antiquity in this region for sealing cracks in ceramic vessels and to fasten tool heads onto wooden handles. Contamination of charcoal samples from Jarmo in Iraq and Jericho in Jordan with petroleum compounds has been suggested as an explanation of apparent anomalies in some of the ^{14}C determinations from these sites.

Organic solvents also need to be employed in attempting to remove waxes, varnishes, shellac, or other coatings placed on samples as preservatives. Unfortunately, records associated with samples stored for long periods of time in museums or other depositories are sometimes incomplete and laboratories must exercise care in dealing with such samples to ensure the unrecorded preservatives do not contaminate samples. It is important that submitters communicate any suspicion or specific knowledge that preservatives were used in the laboratory processing of a sample.

Although the use of organic solvents has been successfully employed in ^{14}C work, their application must be carefully evaluated since all solvents contain carbon. The removal of solvents can be difficult, particularly where sample surfaces are porous and the solvent has been deeply absorbed. In some cases, the removal process requires a series of steps. Unless the submitter of a sample has extensive experience in these procedures, treatment of samples with solvents should be left in the hands of ^{14}C laboratory personnel.

3.3 SAMPLE MATERIALS

During the first decade of ^{14}C studies, certain traditions developed about the general reliability of various types of sample material. Reliability was gauged largely in terms of the assumed resistance of a sample to the influence of external contamination. Initially, Libby (1952a:43) recommended sample materials in the following order: (i) charcoal or charred organic materials such as "heavily burned bone," (ii) well-preserved wood, (iii) grasses, cloth, and peat, (iv) well-preserved antler and similar hairy structures, and (v) well-preserved shell. The "heavily burned bone" organics referred to were carbonized skin and tissue material. Bone, as such, was considered a poor risk. Shell was placed at the bottom of the list because of the possibility of isotopic exchange with groundwater. In the early 1960s, Edwin Olson (1963) reviewed the ^{14}C literature to catalog the degree of acceptability of various sample materials. His conclusion was that wood was unquestionably the best material to use. Charcoal was excellent if restricted to macrosample size pieces. Marine shell generally yielded accurate values if appropriate precautions were exercised. In his view, bone and antler were to be avoided.

More than thirty years of ^{14}C studies have witnessed significant expansion in the variety of sample types employed as well as in refinements in pretreatment and analytical procedures applied to samples (Evin, 1983; Olsson, 1983b). However, despite these developments, the vast majority of ^{14}C determinations continue to be made on charcoal and wood (cf. Walker *et al.*, 1983). In archaeological applications, increasing attention has been focused on the relationship of sample to environmental and cultural context. For example, bone and shell ^{14}C determinations obtained on species widely used in a particular region for food would generally have a higher probability of being directly associated with cultural materials found in the same level in a site than individual charcoal fragments. Also, ritual paraphernalia or heirloom items would not necessarily reflect the age of the deposit from which they had been recovered and would also date "too old." Thus, each sample type has cultural/contextual as well as biophysical/biochemical characteristics insofar as its usage for ^{14}C analysis is concerned (Section 5.2).

3.3.1 Wood and Wood Derivatives (Charcoal)

The largest percentage of samples to which ^{14}C dating has been applied comprises plant materials derived from trees, whether wood itself (twig fragments to large beams) or charred or carbonized wood (charcoal). From the beginning of ^{14}C studies, wood and charcoal, from a physical point of view, were recognized as excellent sample materials because of their very high molecular weight components, allowing rigorous pretreatment procedures to be employed to extract potential contaminants without significantly reducing the amount of indigenous sample remaining. Of whatever specific form, whether roots, stems, or leaves from trees, shrubs, or woody vines, all wood and charcoal samples are derived from plants structurally characterized by cellular systems and chemically characterized by the presence of cellulose, carbohydrates, and lignin materials. Different types of plants, however, vary in their internal organization. The most familiar structure is that of the woody gymnosperm and dicotyledonous plants that are built up with a series of specialized tissues beginning with the surface bark, cambium, sapwood, and the internal heartwood. These plants increase in size as a result of the periodic deposition of concentric layers of cells—"tree rings"—in the transverse section of the trunk and stem structures. By contrast, monocotyledonous plants such as palms are understood to expand vertically by increasing the number and longitudinal dimensions of primary tissues and not by additional separate layers of new growth (Weisberg and Linick, 1983).

For ^{14}C dating purposes, it is important to consider the internal parts of those types of plants with multiple ring structures. Typically, each ring represents cells deposited during one growing cycle, generally one interval during a 12-month period. These cells isotopically reflect the characteristics of the adjacent atmosphere during the period of growth. Dendrochronological calibration of the ^{14}C time scale is possible because of the existence of yearly samples of atmospheric CO_2 fixed into each ring by photosynthesis. One study using high-activity ^{14}C or bomb ^{14}C as a tracer has concluded that there is little, if any, movement of carbon bearing materials between tree rings once the sapwood tissue has been converted into heartwood at least in the species of trees that have been previously used in calibration studies (Berger, 1970a, 1973). However, another investigator identified the presence of bomb ^{14}C in some chemical extracts of heartwood in recent trees (Olsson, 1980a). Other researchers have noted that different wood fractions may exhibit differences in ^{14}C content since mobile oils and resins in some whole woods can differ in certain cases from the cellulose and lignin fractions by up to several percent. Stringent pretreatment of such wood samples, particularly if they are to be used in ^{14}C/dendrochronological studies, would be necessary (Baxter and Farmer, 1973a; Mc Phail et al., 1983). There seems to be no evidence that initial ^{14}C activities in woods reflect variations in soil environments, as, for example, in calcareous as opposed to high organic soils (Tauber, 1983; cf. Olsson, et al., 1972:268).

Ring structure must also be considered when a wood or charcoal sample is derived from species of trees that are characteristically relatively long lived. The average life span of different species of trees varies greatly—from less than 100 years for some birch and fir and 400–500 years for some species of oaks to several millennia in the case of the sequoia and bristlecone pine. A cross section of a tree used to fabricate a large beam might contain, for example, 250 rings. Wood taken from different parts of this beam would thus exhibit a 250-year range in ^{14}C ages (ignoring the possibility of missing or double rings). The outermost ring would exhibit an age close to the cutting or death date of the tree while the center ring might be interpreted as 250 years "too old." This type of problem was first called the "postsample-growth (or inner wood) error" and then renamed, more appropriately, the "presample-growth error" (Ralph, 1971:4).

Probably the most notable example of the effect of this anomaly was associated with the issue of the correlation of the pre-Hispanic Lowland Maya calendar with the modern Western calendar. At the time of the introduction of the ^{14}C method, most Mayanists had tentatively accepted, from a long list of correlation formulas, the "GMT" (Goodman–Martinez–Thompson correlation) as best fitting the available evidence. The first ^{14}C determination bearing the Maya correlation problem was obtained on a piece of wood extracted from a lintel inscribed with a Maya long count calendar notation from the Guatemalan site of Tikal (Kulp et al., 1951:566). Rather than supporting the GMT correlation, however, the ^{14}C value on the lintel supported one of the correlation schemes proposed by H. J. Spinden. This correlation calculated dates exactly 160 years earlier than did the GMT formula. A second ^{14}C determination on another inscribed wooden lintel from Tikal bearing the same long count data as in the first test also supported the Spinden correlation (Libby, 1954b:740).

These results stimulated an intensive dating program undertaken by several ^{14}C laboratories on wood samples taken from a number of Lowland Maya sites. In the critical interpretation of the resultant ^{14}C data, the presample-growth factor was employed to explain the earlier determinations (Satterthwaite and Ralph, 1960; Ralph, 1965). New collections of wood samples with due consideration of this problem resulted in more than 100

^{14}C values that provided strong, but to some, not conclusive evidence for the GMT correlation (Nicholson, 1977; cf. Andrews, 1978; Kelly, 1983). The presample growth factor was also invoked to resolve problems or correlating ^{14}C determinations on charcoal with the apparently well-known chronological sequence for the Classic period at Teotihuacan in Mexico (Kovar, 1966). In the Old World, an example of the effect of this factor is evidenced in the interpretations of ^{14}C determinations on wood associated with a royal residence built during the reign of Charlemagne (early A.D. 800s). One sample (Gif-1820) yielded an age of 1350 ± 100 or about A.D. 600. Since this period does not exhibit significant de Vries effect variability, the 200 ± 100 year anomaly was explained as an age effect of the wood used (Delibrias *et al.,* 1974).

When the ring structure of a wood or charcoal specimen is intact, it is sometimes possible to infer the relationship of a given sample to the center of the original tree structure and thus provide an estimate of the potential amount of presample growth error. In many cases, this is not possible. The problem of using such "long-lived" samples should be carefully evaluated especially when other ^{14}C determinations on "short-lived" samples such as reed or seeds are also being obtained. One should also be alert to the fact that bark represents a composite sample since it contains more growth increments than an equivalent thickness of sapwood or heartwood. Initially, there was concern about the possible effect of decay or rotting processes on ^{14}C concentrations in wood. No variation in ^{14}C activity from this cause, however, has ever been documented (Libby, 1955: 116; Broecker *et al.,* 1956:158).

Rootlet intrusion in charcoal samples can pose difficulties especially where the contamination involves dense, fine hair root growth. Several procedures have been employed to remove rootlets, including treatment with sodium hypochlorite (NaOCl), nitric acid (HNO_3), sulfuric acid (H_2SO_4) and acetone (CH_3COCH_3). Unfortunately, sample loss with such treatments can be as much as 40% (Damon *et al.,* 1964; Ogden and Hay, 1965; Haynes, 1966). Different environments and types of wood vary in the degree to which root contamination can involve serious problems (Ogden and Hay, 1973).

Early in the history of ^{14}C studies, much attention was given to the potential contamination of wood and charcoal with soil humics. Charcoal especially can absorb significant amounts of humic acids (Cook 1964). A number of ^{14}C measurements were made on base soluble fractions extracted from these samples (e.g., Olson, 1963:53; Olsson, 1979b, 1980a). In a large percentage of cases, extracted humic materials exhibited no statistically significant difference in age when compared to the ^{14}C values from the total wood or cellulose extracted from the wood. Under some conditions, however, variations of as much as 1000 years have been reported on a small number of Holocene charcoal samples (cf. Goh and Molloy, 1973; Bailey and Lee, 1973). Some variations might be noted, for example, in samples excavated from deposits overlaid with high organic materials such as peat. One instance has been reported where the base-soluble fraction of a sample of charcoal contained bomb ^{14}C since its ^{14}C activity registered in excess of the contemporary standard, whereas the charcoal itself with the base-soluble fraction removed exhibited a ^{14}C age of about 400 years (Polach et al., 1968:196).

The potentially most serious problem, however, would most often come from samples of Pleistocene age. In one case, a humic acid fraction isolated from a wood sample exhibited an infinite ^{14}C age ($>43,000$, L-479, L-479A), while the wood from which the humic acid had been removed by base extraction yielded a finite ^{14}C value ($33,600 \pm 2800$). In another instance, the humic acid exhibited finite ^{14}C ages ($31,800 \pm 2500$ and $23,000 \pm 1500$) whereas the wood following base extraction exhibited an infinite ($>40,500$) ^{14}C age (Olson and Broecker, 1961:144). Clearly, where age values are expected to approach the maximum ^{14}C age range (i.e., counting rates near background values), rigorous pretreatment to extract the base-soluble fractions is highly desirable. In critical wood samples where sample preparation uncertainties must be reduced to a minimum level, multiple ^{14}C analysis of the different fractions obtained in base extractions can be profitably carried out. This approach was taken, for example, in the case of ^{14}C measurements on wood (in the form of twigs) associated with a mastodon slain and butchered by hunters at the Taima-taima Early Man site in Venezuela (Bryan et al., 1978). Three fractions (HCl-treated, NaOH insoluble, and NaOH soluble) all yielded statistically indistinguishable (at the one sigma level) ^{14}C values of about 13,800 ^{14}C years B.P. (Robinson and Trimble, 1981:320–321). Because of its inert characteristics, cellulose has been regarded as the most reliable component of wood to be used in ^{14}C analysis. In some laboratories, when sample sizes permit, this component is routinely isolated from whole wood samples to minimize potential contamination effects (Gupta and Polach, 1985:9). The exclusive use of the cellulose fraction has been advocated for wood samples being used in dendrochronological studies (Jansen, 1970).

3.3.2 Nonwoody Plant Materials and Textiles

Nonwoody plant remains include grass, reed, leaves, grains, seeds, and manufactured products using plant materials such as papyrus, linen, and cotton textiles including those found in a carbonized (burned) state. Such sample materials represent "short-lived" samples—in some cases from a single season—in that they typically have a relatively short growth period and are generally incorporated into a sample matrix, at most, within a few years of death. In such cases, the presample-growth error that might be a problem in the case of wood would not be encountered.

However, short-lived samples exhibit other problems. The ^{14}C content of a typical wood or charcoal sample reflects the *composite* ^{14}C activity of the total number of tree-rings making up the sample matrix. Short-term seasonal and annual variations in ^{14}C activity, which can amount to as much as several percent, are "averaged out." By contrast, the ^{14}C activity in short-lived plant materials will typically reflect these seasonal and annual variations. Thus, it is important that these effects be considered when comparing ^{14}C values from wood/charcoal to those from such materials as reed, grains, or seeds. This is particularly true in situations where high-resolution ^{14}C age estimates are desired.

In addition, $\delta^{13}C$ values exhibited by many nonwoody plants may contrast with those typically exhibited in wood/charcoal (Bender, 1971). Many grains and grasses from temperate climates exhibit $\delta^{13}C$ values in the same general range as those of wood, averaging around $-25‰$ with respect to PDB (Section 3.6). These plant materials follow exclusively the C_3 or Calvin photosynthetic pathway. By contrast, it has been determined that a large group of tropical and arid-land plants including, for example, sugar cane, millet, and corn *(Zea mays)* operating with a Hatch–Slack cycle and known as C_4 plants, exhibit $\delta^{13}C$ values clustering around $-12‰$ (Bender, 1968). There is, in addition, a third type of carbon-fixation mechanism known as Crassulacean acid metabolism (CAM). CAM plants may operate as C_3 organisms, but under environmental stress can operate similar to C_4 plants and thus alter their carbon isotope ratios accordingly (Bender, 1968; Troughton, 1972; Bender *et al.*, 1973; cf. Browman, 1981:268–280). These differences in photosynthetic pathways reflected in ^{13}C variations can be reflected in ^{14}C differences between woody and non–woody plants of the same age by as much as several hundred years. Section 4.6 discusses the specific procedures involved in correcting for these variations.

Radiocarbon age estimates can be obtained on paper and other plant-derived materials such as cotton if due attention is given to the nature and sources of the organics comprising such samples. In the case of paper, variability in manufacturing technology and sources of feed materials must be taken into consideration since materials of quite different age (e.g., previously used cotton fabrics or wood pulp from bark) can be incorporated into a paper product. The measured ^{14}C age can antedate the date of manufacture sometimes by a significant margin, e.g., up to several centuries. Thus, generally an accurate assignment of age for these types of materials often must rely more on detailed physical and chemical examination than on ^{14}C data (Burleigh and Baynes-Cope, 1983).

It should also be noted that aquatic plants growing in freshwater (spring) environments, in the same manner as some freshwater shells (Section 2.3.4), can exhibit apparent ages of many thousands of years (e.g., Hakansson, 1979; Long and Muller, 1981:192). In areas with a number of artesian springs where local populations were known to have exploited the aquatic plants as a food resource, it would be prudent to be aware of this potential problem.

3.3.3 Marine Shell

Shells are the product of secretions produced by the mantle in all mollusca. In most cases, the shell consists of an organic matrix (conchiolin) impregnated with inorganic salts, predominately calcium carbonate but also including small amounts of magnesium carbonate and calcium phosphate in the form of an aragonite or calcite. In contemporary marine shell, the organic fraction comprises only a few percent of the total shell weight. Thus, almost all ^{14}C determinations on shell have employed the inorganic carbonate fraction. The initial ^{14}C activity in the carbonates of a marine shell reflects the ^{14}C activity in the adjacent ocean water at the time a shell crystallizes its initial structure.

The accuracy of ^{14}C determinations on inorganic carbonates from marine shells was debated in the early years of ^{14}C studies (Kulp et al., 1951; Anderson and Libby, 1951; Rafter, 1955b; cf. Olson, 1963). For example, charcoal and marine shell samples of presumed equal age based on stratigraphic associations from several archaeological sites in Peru exhibited ^{14}C values that varied by as much as 800 years (e.g., Johnson, 1955:150; Kulp et al., 1952:409). It was assumed that the marine shell was yielding the grossly erroneous values. Several reasons for this anomaly were advanced. The first concern was that, being in an inorganic form, the carbonate matrix would easily be exchanged with groundwater carbonates (Kulp et al., 1952). It was also quickly determined that the $\delta^{13}C$ values of marine shell carbonates varied from those of terrestrial organics due to natural fractionation effects (Section 4.3.2). It was thought that this alone should inject a variation of up to 400 years in the ^{14}C age on marine shell when compared to terrestrial organics.

Several avenues of inquiry were pursued to determine more precisely the nature of the problems surrounding the use of marine shell. One approach measured terrestrial organic/marine shell sample pairs in cases where it could be reasonably assumed that they were deposited at about the same time in geological or archaeological contexts. As a number of additional paired values became available, it seemed that, except for several relatively localized areas, the ^{14}C values of the carbonate fraction of well-preserved marine shells differed from standard terrestrial materials (e.g., wood and charcoal) generally by no more than a few percent—*ca.* 100–300 years (Blau *et al.,* 1953; de Vries and Barendsen, 1954; Suess, 1954b; Rubin and Suess, 1955; Dyke and Fyles, 1962; Olson and Broecker, 1959; Hubbs *et al.,* 1960; Fergusson and Libby, 1963).

"Well-preserved shells" are those in which the outer surfaces of the shell have remained intact and have not taken on a powdery or "chalky" appearance. Where shells exhibit such characteristics, the removal of the surface layers with dilute acid can be employed to remove the portion of the shell matrix suspected of having been affected by exchange reactions. In some shells, these exchange reactions are associated with a recrystallization process. In the majority of situations, the primary aragonite, the principal component of shell carbonate in most species, can be dissolved and the reprecipitated carbonate exhibits a calcite structure with a different isotopic composition, as influenced by exchange with bicarbonates in the immediate environment. The degree to which this has occurred can be monitored by several techniques including x-ray diffraction and infrared spectroscopy (Grant-Taylor, 1972). Where a shell exhibits physical evidence of having been recrystallized, especially if it is expected to be of Pleistocene age, it is helpful to obtain such analytical data to determine the degree of recrystallization (Vita-Finzi, 1980:766–767). However, it is sometimes difficult to monitor recrystallization processes on a quantitative basis (Goslar and Pazdur, 1985).

With increasing experience, it was generally accepted that with appropriate corrections, most Holocene-age marine shell carbonate ^{14}C values can be employed to obtain reasonably reliable age estimates. What had to be considered, however, were reservoir effects. In the case of marine shell, the principal reservoir correction involves an "upwelling" phenomenon in which the zero-age ^{14}C concentration of marine shell has been affected by carbonates "upwelled" from the deeper parts of the ocean. Such carbonates are commonly depleted in their ^{14}C concentration by relatively long residence times in the deep ocean. Marine shell ^{14}C activity reflects the degree of mixing between the "upwelled" and surface carbonates. Specific procedures used to correct ^{14}C determinations on marine shells for fractionation and reservoir effects are reviewed in Chapter 5.

With many samples of Pleistocene age, the ^{14}C age of the shell carbonates tends to be somewhat older than the corresponding terrestrial organics. This suggests that surface exchange in samples older than about 10,000 years may effect the age of the carbonate fraction. Such results have led to the work of several laboratories that have dated different fractions of the same shell sample. It appears that the effects of groundwater exchange are typically limited to about the outer 25% fraction of the shell (Olson and Broecker, 1961; Dyck and Fyles, 1964, 1965; Dyck *et al.,* 1966; Lowdon *et al.,* 1967; Lowden and Blake, 1968, 1970; Taylor, 1970). In the preparation of such marine shell samples, removal of 25–50% by weight of the shell is a typical practice when sample sizes permit. In one laboratory, both an outer and inner fraction of a shell is dated when the sample is expected to be of significant age (>30,000 years). However, even if the two fractions exhibit essentially equal ^{14}C ages, the reliability of the age assignment is still often questioned (Nydal *et al.*, 1985).

The examination of the organic or conchiolin fraction of marine shell has been conducted by several laboratories with a degree of uncertainty in the results obtained. Data published by Berger *et al.* (1964) suggested that the carbonate and conchiolin fractions exhibited essentially identical ^{14}C values when the appropriate reservoir corrections are applied (see Chapter 4). Determinations obtained by Taylor and Slota (1979) suggested that the conchiolin fraction might become contaminated by the assimilation of microorganisms. While the effect is not pronounced in material of relatively recent age, the contamination increases as a function of the time that the sample has been subject to attack by soil microorganisms. In extreme cases, the conchiolin fractions can be from approximately 1000 to 2000 years younger than the carbonate fraction. By contrast, Masters and Bada (1980) report that a ^{14}C determination on the total amino acids from the organic fraction of a shell was about 5000 years *older* than the carbonate fraction. They interpret their results to suggest that the carbonate was yielding anomalous results. Clearly, additional investigations need to be carried out to determine the exact nature of the discrepancy. Whether the problem lies in sample preparation or in some systematic biochemical anomaly is not yet clear.

Further difficulties are sometimes reported on shell materials that may relate to problems in the type of shell and location of collection. For example, barnacle shells adhering to a limestone substratum collected from Totthest Island, Australia, yield an age of 13,700 ± 130 (Y-332) and 5180 ± 100 (Y-333) in a context where the expected age was about 2000 years. The tentative explanation offered was that the shell utilized limestone to build its shell or the sample included limestone that adhered to the shell structure (Deevey *et al.*, 1959:156). In another example from a geologic context, shell fragments of different ages came to rest in apparent stratigraphic association in conglomerate deposits. Until this fact had become recognized, what were considered to be seriously anomalous ^{14}C values ranging over tens of thousands of years had been obtained from the mixed deposits (Panin *et al.*, 1983). These situations would probably be encountered in strictly archaeological contexts only rarely, but it would be profitable to be forewarned and alert to such possibilities with shell materials.

It appears that the carbonate fraction from most well-preserved Holocene marine shell collected from an open ocean environment can yield

[14]C values that are as accurate as standard terrestrial values if two conditions are met: (i) reservoir effects have been studied, and (ii) stable carbon isotope values are available. However, since in some regions there tends to be greater natural variation in the ^{14}C content of marine carbonates of equal age, the precision of the marine shell ^{14}C values in such regions may be inherently limited to something on the order of ±200–400 years (cf. Broecker and Olsen, 1959). In some areas (e.g., the western Pacific around Australia and New Zealand), the inherent variability appears to be somewhat less. For Pleistocene-age marine shell, geophysical and postdepositional effects can introduce significant variability in apparent ^{14}C activity. Upwelling effects operating during the Pleistocene for many areas are not well documented (cf. Molina-Cruz, 1977). Postdepositional contamination of some samples through recrystallization of the carbonate material can become severe. Careful attention to sample selection and strict attention to pretreatment procedures are absolutely mandatory for such marine shell samples (Chappell and Polach, 1972).

3.3.4   Terrestrial Shell and Eggshell

Radiocarbon age estimates obtained from shells derived from nonmarine environments (land snail shells or freshwater gastropods) are among the least suitable samples for ^{14}C analysis for archaeological applications. Although marine and terrestrial shells fix carbon in similar ways, the initial ^{14}C levels in many terrestrial environments can vary significantly. This is due to the fact that ^{14}C activities in the sources of carbon utilized by nonmarine shells may not be in equilibrium with atmospheric [14]C. Early studies identified several pathways by which carbon of varying ^{14}C activity might find its way into freshwater shell. It was argued that this would result in apparent ages for terrestrial shells ranging up to several thousand years (Rubin et al., 1963; Keith and Anderson, 1963; Rubin and Taylor, 1963; Tamers, 1970). Goodfriend and Hood (1983) studied three sources of carbon that can contribute to land shell carbonates: CO_2 released from plant respiration, atmospheric CO_2, and CO_2 released from limestone. The most serious problem was assumed to derive from a "limestone effect" due to the dilution of ^{14}C activities in shell by exchange with carbonates distributed in groundwater containing no [14]C.

Measurements of ^{14}C activities in modern freshwater shell of known collection date suggest that typical terrestrial reservoir ages are on the order of several hundreds rather than thousands of years (Evin et al., 1980; Burleigh and Kerney, 1982; Preece et al., 1983; Burleigh, 1983). Studies conducted on samples from several archaeological sites in France noted that snail shell exhibited from about 300–1300 ^{14}C years in excess of the age determinations on associated organics. However, modern gastopods from some environments, such as artesian springs, can yield apparent ^{14}C values in the range of 20,000–30,000 years (Long and Muller, 1981:206; Riggs, 1984). While nonmarine shells can in some cases be employed to provide broad-scaled estimates of age (e.g., to distinguish late Pleistocene from early Holocene contexts), for most situations of interest to archaeologists, their use should probably be restricted to situations where no other more suitable sample types are available and detailed studies of reservoir effects can be or have been undertaken (e.g., Carmi *et al.*, 1985; Sheppard *et al.*, 1986). If possible, the ^{14}C activities of the organic or conchiolin fraction of the shell can also be measured to evaluate exchange, reservoir, and fractionation effects.

By contrast with terrestrial snail shells, controlled studies of the sources of carbon exhibited in avian eggshells indicate that the ^{14}C activity in both the carbonate and organic fractions seem generally to be in equilibrium with the contemporary food source and atmospheric CO_2 (Long *et al.*, 1983). These experiments were designed to evaluate the possibility that reduced initial ^{14}C activities in eggshell samples would result from the "limestone effect," i.e., the intake of limestone fragments by the animal during normal feeding activity. Fortunately, there seems to be little effect on the ^{14}C content of either component of the eggshell when high $CaCO_3$ content feeds were ingested. The conclusion of this study was that age inferences based on the ^{14}C analysis of fossil eggshell should generally yield reliable results if appropriate attention to detail in sample preparation is observed, e.g., careful removal of 50% or more of the outer layers of the shell to reduce the possible effect of postdepositional exchange.

3.3.5 Bone and Related Materials

Bone is a specialized form of calcified tissue forming the skeletal framework of the bodies of vertebrates. Carbon containing compounds exist in bone tissue in both *inorganic* and *organic* forms. The proportion of each varies among different species as well as between different bones and bone structures within the same animal. Generally speaking, however, the inorganic fraction makes up about two-thirds of most fresh, dry, compact bones in the form of apatite (calcium phosphate crystals with the structure of a hydroxyapatite), calcium carbonate, and other amorphous inorganic materials. The organic component of fresh, fat-free bone is composed largely of the protein collagen deposited in a dense network of laminated fibers. The physical structure of bone has been compared to a brick wall with the apatite crystals as bricks, a mortar made up of a complex "ground substance" (mucopolysaccharides, glycoproteins, lipids, carbonate, citrate, sodium, magnesium, and a host of trace components) and with collagen fibers acting as reinforcing steel rods (Berger *et al.*, 1964).

Early in the development of the ^{14}C method, bone (like shell) acquired a reputation as an unreliable sample type (Libby, 1952a:43; de Vries and Barendsen, 1954; Ralph, 1959:56; cf. Olson, 1963:61–65). (It should be noted that what was labeled as *burned bone* was generally highly regarded. In this case, however, the actual samples were derived largely from the carbonized skin and tissue and not from the bone structure itself.) It was assumed that because of its relatively porous structure and largely inorganic carbon content, there would be a high probability of isotopic exchange with groundwater carbonates. Early measurements obtained on whole bone confirmed suspicions that, generally speaking, it could not be trusted (Sinex and Faris, 1959). It was quickly shown that the ^{14}C activity in the inorganic or carbonate fraction was indeed often anomalous (Broecker and Olson, 1961:142). There was the fear that even an organic fraction of bone might be contaminated by humic materials dissolved in groundwater (Munnich, 1957). For many sites, the availability of usually reliable sample types such as charcoal, initially obviated the need to employ such problematical samples. Where stratigraphic discontinuities created a question of association with standard samples, or where no other sample could be recovered, there continued to be an interest in determining if some portion of a bone sample had the isotopic integrity required to yield accurate ^{14}C values.

Table 3.1 lists the different inorganic and organic fractions that have been employed in obtaining ^{14}C determinations on bone. In the majority of samples derived from typical archaeological or geological environments (in contrast to fresh bone), the inorganic fraction is composed both of the original apatite component and diagenetic or secondary carbonate. The ^{14}C activity of the secondary carbonate fraction generally reflects the

TABLE 3.1
Major Fractions of Bone Used for ^{14}C Determinations

I. Inorganic fractions
    A. Indigenous carbonates in apatite structures
    B. Diagenetic/secondary carbonates
II. Organic fractions
    A. Acid soluble, acid insoluble, base soluble, and base insoluble fractions (total and specific molecular weight ranges)
    B. Collagen
    C. Gelatin
    D. Total amino acids
    E. Specific amino acid (e.g. hydroxyproline)

degree of isotopic exchange with groundwater carbonates rather than the actual age of the bone. Radiocarbon values obtained on the total carbonate fraction can be older or younger than organic fractions from the same bone sample. An example of a carbonate fraction significantly older than the organic component is exhibited on an artifact that played an important role in discussions concerning the timing of the initial occupation of the Western Hemisphere. An age of $27,000^{+3000}_{-2000}$ ^{14}C years B.P. (GX-1640) was initially assigned to an apparent tool made from a caribou tibia from an Old Crow Basin locality in the Yukon Territory, Canada. The ^{14}C determination was made on a sample of CO_2 released from the bone by treatment with acid (Irving and Harrington, 1973). Although this fraction was characterized as "bone mineral apatite," a more appropriate designation would have been "total carbonate fraction." A subsequent ^{14}C analysis by AMS techniques (Section 4.5) determined that the ^{14}C age of an organic fraction of this bone (RIDDL-145) was 1350 ± 150 ^{14}C years B.P. (Nelson et al., 1986a). On the other hand, ^{14}C analyses of an organic and inorganic fraction of a terminal Pleistocene bison tibia from the 12 Mile Creek site in Kansas (Rogers and Martin, 1984) yielded concordant values: $10,435 \pm 260$ ^{14}C years (GX-5812A) on the inorganic fraction and $10,245 \pm 335$ (GX-5812) on an organic fraction. It is clear that it is difficult to generalize concerning the age relationships of the organic and inorganic fractions in bone without detailed geochemical information concerning the depositional environment (McPhail, 1982; Donahue et al., 1984; Stafford et al., 1984).

Experiments aimed at isolating the carbonates from the actual apatite component in bone unfortunately have yielded inconsistent results. Under certain conditions, ^{14}C analysis of this fraction seems to yield accurate values (Haynes, 1968). However, geochemical and mineralogical studies have revealed a number of mechanisms that can significantly alter the carbon isotope values in apatite structures (Hassan, 1976; Hassan et al., 1977). Such obstacles may not totally discourage attempts to use the apatite fraction, as other workers have reported more encouraging results using thermal decomposition to isolate CO_2 from the in situ apatite structure (Hass and Banewics, 1980).

The questionable reliability of ^{14}C determinations on inorganic components of bone has led the majority of researchers to concentrate on one or more of the organic fractions. All chemical pretreatments assume an initial physical examination of the external surface and fracture zones to insure the removal of preservatives, microorganisms, rootlets and other nonbone organic fragments (Hassan and Ortmer, 1977). Chemical processing involves initially the elimination of all inorganic carbonates. Both EDTA and HCl have been used for this purpose (Berger et al., 1964; Olsson et al., 1974). However, fear of contamination with "old carbon" from the EDTA treatment has been expressed. Such a problem can apparently be minimized or eliminated with sufficient washing (Fakid *et al.*, 1978). Further preparations have included conversion to gelatin (Sinex and Faris, 1959; Longin, 1971; Berglund *et al.*, 1976), treatment with NaOH to remove humates and other base soluble fractions (Olsson *et al.*, 1974; Haynes, 1967), and the separation of total amino acids (Ho *et al.*, 1969; Taylor and Slota, 1979). The organic fraction that would be least likely to be contaminated would be a single amino acid such as hydroxyproline, known to have limited distribution except as a constituent amino acid of collagen. Until recently, the use of an amino acid fraction on a routine basis has usually been impractical because of sample size limitations. However, with the advent of AMS ion counting (Section 4.5), this approach now is becoming increasingly feasible and is becoming the method of choice for critical age assignments (Wand, 1981; Stafford *et al.*, 1982; Taylor *et al.*, 1984a; Gillespie *et al.*, 1984b).

Table 3.2 illustrates the excellent degree of concordance between three types of organic samples, including a bone organic fraction from a single burial feature (part A), as well as the range in ^{14}C values on different fractions of two bones from the same Holocene age skeleton (part B). In the second example, the total carbonate values are significantly younger than the ^{14}C determinations obtained on the organic fractions. By contrast, the acid-insoluble and total amino acid fractions exhibit essentially identical ^{14}C values and indicate an age consistent with the cultural affiliations of the archaeological materials associated with the burial. In this case, the preservation of the organics in this skeleton was very good.

The goal of all pretreatment procedures for bone is to isolate one (or more) organic fraction(s) which is (are) unambiguously indigenous to the original sample. Various criteria have been employed in an attempt to identify those bone samples that may have been contaminated. One quantitative approach has been called the "pseudomorph" test. Figure 3.1 illustrates where demineralization of bone from a late Roman/Christian cemetery in England has left behind an organic replica or pseudomorph of the original bone structure. The same pretreatment of a bone sample from a Pleistocene site in North Africa yields a product in which the original physical structure is totally lost (Fig. 3.2). Unfortunately, a high degree of physical degradation in the bone matrix may or may not correlate with the presence of significant amounts of contamination.

A more promising approach has been to use amino acid composition as a means of characterizing the organic constituents of bone (Taylor, 1980; Hassan and Hare, 1978). Collagen in modern mammalian bone is characterized by a high glycine content, relatively high proline, and, as noted before, hydroxyproline. This amino acid signature can be employed to determine the degree to which the organic fraction extracted from a

TABLE 3.2

Radiocarbon Determinations on Human Bone and Associated Samples

Sample number	Sample/fraction	Organic carbon yield (%)	$\delta^{13}C$ (‰)[a]	Conventional ^{14}C age (^{14}C yr B.P.)
A. "Whiskey Lil," Chimney Cave, Nevada (Berger *et al.*, 1965)				
UCLA-692	Cedar matting	—	—	2590±80
UCLA-690	Skin	—	—	2510±80
UCLA-689	Bone (total acid insoluble)	—	—	2500±80
B. Burial 36, CA-SJo-112, California (Taylor and Slota, 1979)				
(i) Rib				
UCR-449A	Inorganic (total carbonates)	—	−8.42	930±140
UCR-449B	Organic (total acid insoluble)	6.95	−19.89	2765±155
UCR-449C	Organic (total amino acids)	1.26	−21.41	2930±150
(ii) Tibia				
UCR-450A	Inorganic (total carbonates)	—	−9.43	830±100
UCR-450B	Organic (total acid insoluble)	6.49	−20.24	2835±140
UCR-450C	Organic (total amino acids)	1.12	−21.29	2960±140

[a] With respect to PDB.

bone sample can be characterized biochemically as collagen. In the case of most bones derived from the typical geological or archaeological site, it appears that one usually is not dealing with *unaltered* collagen (Tuross and Hare, 1978). Because of this, it is probably inappropriate to label an organic extract from an archaeological bone as "collagen" unless a biochemical analysis on the organic fraction being dated has been obtained. It probably is preferable to use such terms as acid soluble, acid insoluble, base insoluble, or gelatin fraction to describe such samples. Figure 3.3 outlines the steps employed in the University of California, Riverside, ^{14}C laboratory to prepare different fractions of bone for ^{14}C analysis based on differential solubility characteristics. Depending on the amount of bone

Figure 3.1 Organic residue following demineralization of human bone sample from late Roman Christian cemetery near Poundbury, Dorchester, England. (Sample courtesy of Theya Molleson, British Museum.)

Figure 3.2 Organic residue following demineralization of human bone sample from Upper Paleolithic site in North Africa.

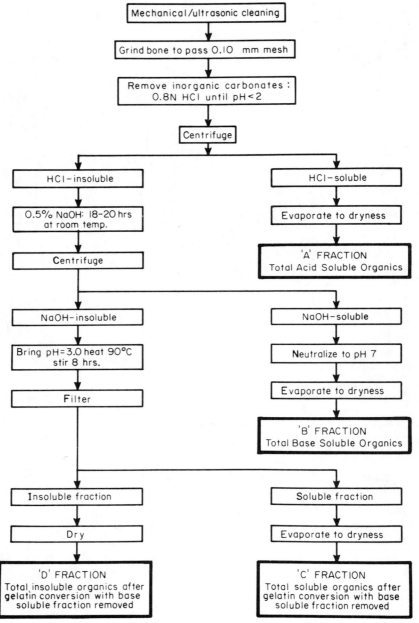

Figure 3.3 Preparation of organic fractions of bone for ^{14}C measurements on basis of differential solubility characteristics. [From Taylor (1983).]

available, up to four fractions are prepared. The degree of concordance in the ^{14}C activity of the fractions is used as one indication of the accuracy of the ^{14}C age estimates obtained (Taylor, 1983).

With appropriate attention to detail in the pretreatment procedures, accurate estimates of the age of Holocene bones can usually be obtained using one or more organic fractions (Taylor and Slota, 1979; Taylor, 1982). Variations in the ^{14}C values among the various organic fractions for such bones are typically on the order of several hundred years. However, in some cases, there can be variations of up to several thousand years. The dating of Pleistocene bone samples requires a much greater attention to the specifics of sample pretreatment because of generally low organic carbon content with the resulting potential for greatly increased effects of very small amounts of modern contamination (Taylor, 1980). Variability among organic fractions in these samples appears rarely to exceed ± 1000–2000 years although there are reported instances of much larger deviations (Horvatincic *et al.*, 1983). For these samples, the advent of ion counting will permit milligram amounts of organic extracts including the use of a single amino acid such as hydroxyproline. The results of these measurements should be in the vast majority of cases comparable to those obtained from standard terrestrial samples such as charcoal or wood. It might also be noted that many of the same strictures associated with bone apply to antler. Organic fractions of the compact portion of antler with appropriate attention to pretreatment can usually yield acceptable values.

3.3.6   Other Sample Types

In principle, ^{14}C analyses can be made on any carbon containing compound. However, 80–90% of all ^{14}C age estimates *on archaeologically related materials* have employed "standard" sample types: wood/charcoal or marine shell. The remaining 10–20% is made up almost entirely of the other sample types discussed in the previous sections. The dominance of these materials is due in large part to the fact that they constitute the most widely distributed and best preserved organics in the majority of archaeological sites. Attempts to develop procedures to obtain ^{14}C estimates on nonstandard sample types usually have been initiated in situations where a standard sample type was not available, not available in sufficient quantity, or where the association of a standard sample type with an archaeological context was problematic to some degree. The ability to obtain ^{14}C determinations on milligram and microgram amounts of carbon using AMS technology (Section 4.5) is creating greatly expanded opportunities for investigations on samples heretofore not practical to examine because of sample size limitations. The AMS ^{14}C analysis of blood residues recovered from stone artifacts (Nelson *et al.*, 1986) provides an excellent illustration of the kinds of nonstandard sample types on which direct ^{14}C analysis can now be applied.

Radiocarbon activity of soil organic fractions is extremely variable and the usefulness of using such values to infer age in archaeological applications is generally quite limited except under special conditions. The basic problem with soil ^{14}C dating is to understand the processes involved in soil genesis and particularly the initial ^{14}C activities in the sources of the organics in the various fractions that can be isolated (Geyh *et al.*, 1983). The degree of concordance of apparent ^{14}C ages on different soil organic fractions prepared from the same bulk soil sample by sequential base extractions can sometimes be used as an index of the overall reliability of the ^{14}C age estimates for the soil horizon (Kigoshi *et al.*, 1980; Gilet-Blein *et al.*, 1980). Difficulties increase with attempts to date Pleistocene soil horizons (Goh *et al.*, 1977). Similar problems are often encountered in working with various types of carbonate deposits such as caliche or tufa. Various sources of significant error are possible from the incorporation of old carbon in detrital form from bedrock sources as well as from isotopic exchange. The validity of dates from such deposits is extremely variable and requires a detailed analysis of control samples before one can employ such ^{14}C values to infer actual age (Srdoc *et al.*, 1980, 1983; Karrow *et al.*, 1984).

A good example of the problem of using ^{14}C values on carbonate encrustations on samples to infer the age of such samples is shown by studies conducted on a human burial excavated in the Yuha desert region of interior southern California. Attention was drawn to this burial as a result of a ^{14}C measurement of 21,500 ± 1000 ^{14}C years B.P. (GX-2674) on caliche coating one of the bones and 22,125 ± 400 (UCLA-2600/1854) on caliche coating a cairn boulder placed over the skeleton (Bischoff *et al.*, 1976). The use of these values to assign a late Pleistocene age to the Yuha burial was challenged on the basis of several lines of evidence (Payen *et al.*, 1978; Wilke, 1978). Radiocarbon determinations obtained by AMS techniques (Section 4.5) on several bone organic fractions of the surviving small fragments of the Yuha human bone produced ^{14}C values of less than 4000 years (Stafford *et al.*, 1984). Another example of this problem is provided from ^{14}C data obtained on the Otovalo human skeleton unearthed from a firmly carbonate-cemented river terrace deposit in the Rio Ambi Valley, Ecuador. Radiocarbon analysis of various types of carbonate deposits removed from several bones ranged in age from *ca.* 28,000 to 35,000 ^{14}C years B.P., whereas analysis of two bone organic fractions yielded values of between 2300 and 2670 ^{14}C years B.P. (Brothwell and Burleigh, 1977).

Trace amounts of organics contained in sherd fragments have been extracted in an effort to provide a direct ^{14}C age estimate for ceramic samples. Initially, it was assumed that any organics present in such samples would be derived primarily from two sources: organics added as temper (e.g. straw) that had survived the firing process or, more likely, organics that had been absorbed into an inside or outside surface during usage, e.g., soot or food residues (Strukenrath, 1963; Engstrand, 1965). It was assumed that the materials used for temper would be essentially contemporary with the pottery fabrication process. Reasonable concordance between ^{14}C determinations on the organics extracted from pottery and stratigraphically associated samples such as charcoal indicated that these assumptions were correct for some ceramics (Taylor and Berger, 1968). However, under certain conditions, discordant results have been documented. The source of these anomalies includes organics contained in the original clay sources of significant age not removed during firing, the use of freshwater and fossil marine shell for temper, the use of fossil fuels (e.g., coal) for cooking, and finally, postdepositional contamination from modern high-organic-content soils (De Atley, 1980). Because of the small amount of organic carbon typically contained in sherds, large sample sizes were required for conventional ^{14}C decay counting; usually kilogram amounts of sherds were needed. With the use of AMS methods of counting (Section 4.5), this problem can be overcome. With appropriate precautions, the ^{14}C analysis of organics contained in sherds and other low-carbon-content anthropogenetic samples (e.g., wattle-and-daub house construction materials) can provide important chronological information in some archaeological contexts.

Small amounts of organic carbon contained in iron can also be extracted to provide the basis of inferring the date of manufacture for iron implements and weapons (Van der Merwe and Stuiver, 1968; Van der Merwe, 1969). In traditional smelting processes, carbon in the form of either wood charcoal or coke (coal) was incorporated to increase tensile properties. The use of coal as a carbon source, however, became widespread only within the last 200–300 years. Thus, in premodern smelting operations, charcoal from freshly cut wood provides the principal source of carbon, and ^{14}C determinations on such samples can be used to assign age to an iron artifact. Traditional methods of iron metallurgy in the Near East, China, Europe, and Africa produced various types of iron/steel alloys with carbon contents ranging from about 0.1 to 5%.

An additional example of a specialized application involves attempts to use the ^{14}C activity in lime-based mortars to assign age to construction activity (Labeyrie and Delibrias, 1964; Delibrias and Labeyrie, 1965). Such mortars were prepared by first heating limestone ($CaCO_3$), which drives off the CO_2, to yield lime or quicklime (CaO). In the preparation of a mortar, the lime is mixed with an aggregate matrix (sand, gravel, or other solids to add bulk), and water is added. As this mixture hardens or "sets up," CO_2 from the atmosphere is absorbed. If it is assumed that the only source of carbon in a mortar sample derived from the CO_2 absorbed during the curing process, then the ^{14}C activity should accurately reflect the age of the construction event associated with the mortar. Unfortunately, discrepancies between the historically documented and apparent ^{14}C ages for a number of mortars have been noted (Stuiver and Smith, 1965; Baxter and Walton, 1970). In these cases, the discordant ^{14}C ages were in excess—sometimes by as much as 2000–4000 years—of the known age of the mortar.

Suggested causes for such anomalies include carbonates contained in the sand/gravel aggregate, the incomplete heating of the limestone when it was originally prepared, which allowed residual CO_2 to remain in the lime, the possible use of shell as the source of $CaCO_3$ and, extremely slow curing rates for some mortars. The most serious problem appears to be the presence of geologically old carbonates, e.g., marble or limestone fragments, which have been added to the mortar. These materials would, of course, contain CO_2 of "infinite" [14]C age. Careful, sample pretreatment to eliminate the carbonate rock chips in the mortar matrix has been advocated as one means of dealing with this problem (Folk and Valastro, 1976). This was accomplished, for example, in conducting ^{14}C measurements on mortar samples from pillars in a cathedral in Belgium. The mortar contained geologically old chalk fragments. When removed, the resulting ^{14}C analysis of the mortar yielded age estimates that agreed reasonably well with the expected ages (fifteenth to sixteenth century A.D.). The ^{14}C analysis also identified recent repairs in one of the pillars (Dauchot-Dehon et al., 1983).

3.4 SAMPLE SIZE, SAMPLE COLLECTION, AND DATA CONSIDERATIONS

Several factors enter into an estimate of the amount of sample needed to obtain a ^{14}C determination. A primary consideration is the carbon content of the specific fraction of a sample on which the ^{14}C analysis is to be conducted. In the case of wood and charcoal, this would be most of the total dry sample weight minus any contaminants removed. In some cases, however, it might be desirable to extract a cellulose or lignin fraction from the total wood sample. In the case of bone, the relevant amount involves

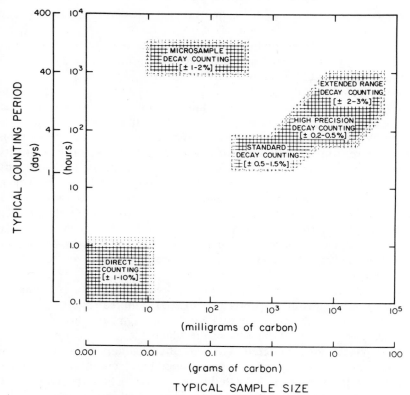

Figure 3.4 Relationship between sample size and typical counting periods for different types of instrumentation employed to measure ^{14}C. Based on Table 4.1.

what *organic* carbon fraction is to be analyzed. For some sample types such as wood and shell, the carbon content is relatively stable. The organic content of charcoal is typically high but can vary depending on the degree of carbonization of the source wood. For other samples, such as bone, the organic carbon content can be highly variable. In these cases, it is usually helpful to obtain an analysis of the nitrogen content so that the amount of organic carbon present can be estimated.

A second important consideration in estimating sample requirements is the type of counting instrumentation to be employed in the analysis. Figure 3.4 summarizes the range in sample sizes typically required with the different types of counting systems used in ^{14}C analysis. (The basis of the differences in these approaches is discussed in Sections 4.4 and 4.5.) The major contrast is in the requirements of direct counting by accelerator mass spectrometry (AMS) and conventional decay counting.

Currently, direct counting for ^{14}C analysis is typically being employed in special circumstances where sample sizes are restricted, e.g., specific organic extracts from bone or single seeds. In addition, statistical precisions are somewhat less than with standard decay counting, but this distinction is disappearing with advances in direct counting technology.

The expected time frame of the sample also has a bearing on the amount of sample to be collected. Generally, samples with expected ages in excess of 10,000–20,000 years require larger amounts of sample material to retain acceptable counting statistics. Extended range decay counting systems often require from 10 to 50 g of carbon for maximum use of the capacity of the instrumentation. Currently, AMS systems are limited in their ability to extend the ^{14}C time frame beyond about 40,000–60,000 years. However, there is a reasonable expectation that advances in AMS technology will permit the extension of the dating range if the exclusion of contamination during sample preparation can be achieved. Sample size requirements to accomplish this may increase from 1–5 mg to as much as 10–20 mg due to the fact that, unlike decay counting, a sample is "used up" during an AMS analysis.

Table 3.3 summarizes typical sample size requirements for (A) standard samples requiring routine pretreatment, (B) samples requiring special treatment, and (C) samples employed only under special circumstances. Samples listed under (A) are those on which the great majority of ^{14}C analyses have been conducted. They are the most amenable to rigorous pretreatment procedures, and in general, laboratories are most experienced in dealing with them. Samples listed in group (B) include those on which extended pretreatment procedures are required (e.g., bone) or where $\delta^{13}C$ values can significantly vary. Group (C) samples are those which previous researchers have examined with varying degrees of success. For many of these samples, the principal problem is the minimal carbon content with the resultant question of how to distinguish *in situ* versus non-*in situ* organics.

Table 3.3 provides a listing of the typical *submission* weights (in grams) of sample material generally needed for two sizes of decay counting and AMS direct counting instrumentation. The two decay counting examples represent upper and lower ranges in sample size requirements for typical decay counting systems. The submission weight is an average amount that needs to be submitted to the laboratory. The amount of sample collected in the field should be significantly in excess of the submission requirements. *It should be emphasized that these figures are intended as only a very general guide*. Individual laboratory requirements can vary significantly depending on changes in equipment and ability to compensate for inadequate sample material. Microsample decay counting allows the use of much smaller samples, whereas counting systems with high

TABLE 3.3

Guide to Sample Size Requirements for Radiocarbon Analysis for Decay and Direct Counting Systems: Typical Submission Weights

Sample type	Examples	Decay counting		Direct counting
		1-liter detector	8-liter detector	
A. Routine pretreatment				
Charcoal		2–5 g	20–40 g	0.5–1 g
Wood		5–10 g	40–60 g	~2 g
Marine shell (carbonates)		10–20 g	50–100 g	10 g
B. Extended pretreatment analysis				
Plant products	Paper, papyrus, textiles, seeds, grains, grass, leaves, coprolites	5–10 g	40–60 g	1–2 g
Animal products	Bone (organics), flesh, skin, hair "burned bone" (carbonized tissue)	100–500 g / 50–300 g	1000–2000 g / 500–2000 g	20–200 g / 10–20 g
C. Special circumstances				
Soil components	Peat, soil organics	10 g	100 g	2 g
Freshwater shell (carbonates)		10–20 g	50–100 g	10 g
Marine shell (organics)		100–500 g	1–3 kg	20–200 g
Bone (inorganic fractions)		20–40 g	100–200 g	20 g
Other animal products	Tusk, antler, ivory, teeth	100–500 g	1–3 kg	20–200 g
Other carbon containing materials	Ceramics, bricks, wattle-and-daub, ceramics (organics) mortars, plaster, iron	1–5 kg	1–3 kg	500 g

precision and/or extended range capabilities usually require larger sample sizes. Laboratories should be contacted before field studies begin to determine current requirements.

The collection of samples in the field has over the years acquired a "lore" that in itself is an interesting example of a scientific oral tradition. Despite early denials by experienced ^{14}C specialists (e.g., Ralph, 1971:809), there remains the view that it is not safe to touch a sample or that samples must be placed in airtight containers immediately upon collection because of the potential for contamination with fallout (cf. *American Antiquity* 18:98, 1952). Both suggestions reflect a salutary emphasis on minimizing the possibility of contaminating a sample with modern organics (including bomb ^{14}C), but they are, except in rare instances, not mandatory. Ordinary cleanliness, caution, and common sense are usually sufficient. One does not want to touch a sample if, for example, one's hand is covered with oil, handcreams, or powdered coal. Routine precautions will avoid placing samples in contact with *any* modern organic materials that might mix with the sample matrix. As an obvious example, this means that samples should not be packed in such materials as cotton or paper cuttings. Clean metal containers with screw tops and metal foil are the best materials in which to package samples intended for ^{14}C analysis—glass containers may break and some types of plastics used to fabricate containers may "outgas" organic compounds and should be avoided. The size of containers should be appropriate to the size of the samples. In addition, one should avoid placing labels in with the sample themselves; labels can be attached to the outside of a container in a manner that will preclude any possibility of being detached (cf. Gillespie, 1984:5).

The data associated with a sample being submitted for ^{14}C analysis should ideally constitute an abstract of the proposed significance of the age estimate(s) that will be obtained by the laboratory. Most ^{14}C laboratories have sample data or submission forms that are intended to be used as guides in recording this information (cf. Sheppard, n.d.:67; Polach and Golson, 1966:35). The categories identified on most submission sheets can be most useful in focusing detailed and specific attention on the nature of samples and their geographical, geological/stratigraphic, and cultural contexts. It would be helpful if such categories were considered as part of the overall research design for the chronological component of an archaeological study before field work is undertaken as well as when samples are being removed from their primary geological or archaeological context during excavation. Although data forms focus on the individual ^{14}C sample, in most cases, an individual sample would be submitted as part of a suite of samples which, taken as a whole, would relate to a specific question or issue being considered (cf. Davis, 1965).

The general categories requested in most data forms typically include the following:

1. *Locality designation:* name of site locality employing standard geographical terminology (county, parish, state, province, UTM coordinates, latitude and longitude, grid reference system, etc.).

2. *Sample materials:* specific identification of nature of the sample organics on which the ^{14}C analysis is to be obtained (e.g., wood, marine shell, bone, dispersed organics in soil, etc.) including genus or species level identification where available. Typically, laboratories will modify this description by noting if a specific chemical fraction is employed (e.g., lignin fraction for wood, amino acid fraction for bone, etc.).

3. *Site type:* identification of primary archaeological context (e.g., house floor, burial, hearth feature, etc.).

4. *General site environment:* local landform (e.g., lacustrine or fluvial deposits, etc.), specific soil type from which sample was recovered, current and, if known, former vegetation cover, and modern land use patterns.

5. *Condition of sample collection zone:* a statement of the physical context of the immediate area from which the sample(s) were taken, e.g., dry, waterlogged, root zone. Proximity to possible sources of contamination, e.g., bogs or other high organic content soils, limestone outcrops, groundwater conditions, presence of petroleum or coal deposits, geothermal springs, volcanism, etc.

6. *Geologic/stratigraphic relationships:* position of sample(s) in stratigraphic context, horizontal and vertical relationship of sample(s) to relevant archaeological features, soil horizons, natural (including surface finds) or excavated context.

7. *Sample treatment by submitter:* description of actions taken by submitter with regard to sample before being sent to the laboratory (e.g., washed, dried, and what other materials were removed from sample matrix) and whether any preservatives were applied.

8. *Cultural significance:* reason for dating this (these) particular sample(s). To what issue or problem does the sample relate? A statement of the relationship to previous ^{14}C values obtained and relevant literature references.

The description of the result of the ^{14}C age estimate as published in data list form in *Radiocarbon* follows this format in somewhat less detail and, in addition, includes the date of collection, name of collector, as well as the name and institutional affiliation of the individual submitting the sample to the laboratory.

CHAPTER 4

MEASUREMENT TECHNIQUES

4.1 NATURE OF RADIOACTIVITY

One of the most fundamental series of discoveries in physics and chemistry was the progressive unfolding of the organization of the physical world as exemplified in the development of the theory of the atom. At the beginning of the nineteenth century, all matter was conceived of as being made up of *elements,* extremely small indivisible spheres or "atoms" (from the Greek word for *indivisible*). Each element was uniquely characterized by a different atomic "weight." By the end of the nineteenth century, it had been suggested that the atom actually consisted of at least two parts: a small positively charged core or *nucleus* and "shells" or "clouds" of negatively charged *electrons* circling the nucleus. In the early 1900s, every element was seen as being characterized in terms of an equal number of *protons* (positively charged particles) in the nucleus and electrons distributed around the nucleus. The nuclear properties of an atom reflecting the composition and structure of its nucleus were represented in terms of the number of protons that defined the *atomic number*.

One important development that led to the conception of this model was the discovery of the phenomenon of *radioactivity*. In 1896, A. Henri Becquerel found that uranium emitted an invisible radiation that in many

71

respects was similar to the x-rays that had been discovered the year before by the German physicist Wilhelm C. Roentgen. Two years later, Pierre and Marie Curie introduced the term "radioactivity" to describe this phenomenon. The fact that certain elements emitted radiation led to the view of Ernest Rutherford and Frederick Soddy that the nuclear properties of some elements were variable. In 1910, Soddy proposed the term *isotope* to designate atoms with an identical atomic number but different *atomic mass*. In the 1930s, it was suggested that differences in atomic mass could be attributable to variations in the number of uncharged particles or *neutrons* in the nucleus. Just as "element" was used for atoms with the same atomic number (i.e., the same number of protons or electrons), the word isotope came to be used to characterize an atomic mass with a unique number of protons and neutrons (Friedlander *et al.*, 1964).

In the case of carbon, for example, its chemical properties using this model are defined in terms of the six protons in its nucleus. Thus carbon has an atomic number of six. The two stable isotopes of carbon, $^{12}_6C$ and $^{13}_6C$, have, respectively, six and seven neutrons in their nucleus. Radiocarbon $^{14}_6C$ contains eight neutrons in its nucleus and is unstable or radioactive. It undergoes *beta decay* in which the mass number remains unchanged, but the atomic number increases by one. This occurs when a neutron changes to a proton with the emission of an uncharged particle (a neutrino) and a beta particle (a negatively charged particle identical to an electron). Thus the number of neutrons decreases by one (from 8 to 7) and the number of protons increases by one (from 6 to 7). This results in the production of a stable isotope of nitrogen, $^{14}_7N$.

Another common natural decay phenomenon involves the production of *alpha particles*. More massive than betas, alpha particles are composed of positively charged helium nuclei (two protons and two neutrons). Alpha radiation most generally occurs in conjunction with the decay of isotopes of relatively high atomic number. One of the alpha emitters commonly found in many ^{14}C sample preparations is radon, ^{222}Rn, produced in the decay series of one of the isotopes of uranium, ^{238}U. Since radon is a gas, it sometimes finds its way into counting gases (Section 4.4.2). Fortunately, since it has a half-life of less than 4 days, storing sample gases for 4–5 weeks usually is sufficient to allow the radon to decay below detection limits. In some samples, however, measurable amounts of radon are present even after 2–3 months (Nydal, 1983).

4.2 METHODS OF MEASUREMENT

With the recognition that an element may have a series of isotopes, physicists set to work to identify and characterize them. Within a decade after the first isotope was discovered, 70 isotopes of 29 elements had been identified. Today, about 280 stable isotopes are known among the more than 100 elements so far identified. Many isotopes do not occur in nature, but have been produced artificially in nuclear accelerators and reactors.

Isotopes can be measured directly through the use of a *mass spectrometer* (Fig. 4.1). As the name implies, these instruments take advantage of the differences in mass of different isotopes to detect and measure their concentrations. The process of measurement requires that the nuclide be *ionized* by stripping off or adding to the electrons on the outer "shells" of the atoms. In these states, the ion can be influenced by magnetic fields. This property of ions permits them to be accelerated in a vacuum. When such acceleration occurs, their trajectories can be deflected when they are passed through a magnetic field of appropriate strength. The degree of deflection of the pathway of a monoenergetic ionized beam largely depends on the differences in mass of the different isotopes comprising it. Figure 4.1 illustrates this principle. The ions of the isotope selected for analysis reach the collector in greatest concentration. Ions of greater mass are insufficiently deflected (because of their greater inertia) for a given magnetic field strength, whereas ions of smaller mass are excessively deflected. These ions lose energy and are removed from the spectrometer by pumping. By varying the strength of the magnet or the energy imparted to the ions, relative concentrations of different isotopes can be measured.

If the amount of a given isotope is sufficient and the differences in mass

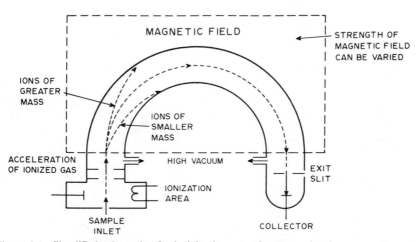

Figure 4.1 Simplified schematic of principle elements of a conventional mass spectrometer. [After Skoog and West (1971:362).]

are large enough, the isotopic composition of a sample can be obtained with a conventional mass spectrometer. This is the method routinely used to obtain the concentration of ^{13}C relative to ^{12}C. Attempts to measure natural ^{14}C concentrations with a conventional mass spectrometer, however, were frustrated because of the difficulties in getting the background of vastly more abundant components of the beam sufficiently reduced to measure natural ^{14}C levels (Anbar, 1978). The development of high-energy or accelerator mass spectrometry (AMS) has made the direct counting of ^{14}C practical (Section 4.5).

Many of the pioneering studies of the physical properties of radioactive isotopes, including ^{14}C, came about as a result of the development and application of very sensitive radiation detection devices. By the late 1930s, these studies included the use of some type of gas-filled counting instrument. A typical instrument of this type is the Geiger tube (Geiger–Müller or GM tube) illustrated schematically in Fig. 4.2. Particles from either outside (as in A to A') or internal (B to B') to the counter can interact with the counter gas to produce ionization. This ionization process creates charged free electrons that are sensitive to the magnetic field set up by applying an electric potential between the central wire and wall of the cylinder. When the central wire is positively charged, the negatively charged electrons are attracted to the wire. As these electrons pass through the gas, they, in turn, produce secondary electrons. Upon reaching the wire, these electrons cause a current to be produced that, when amplified and connected to appropriate instrumentation, can be used to detect decay events. Thus the presence of a radioactive nuclide and its concentration can be inferred by measuring its decay products. In the case of ^{14}C, this

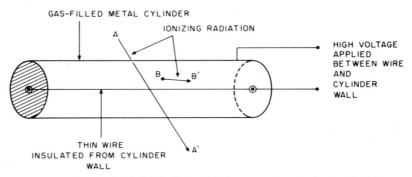

Figure 4.2   Schematic of simple Geiger–Müller type of gas ionization detector.

involves the detection of the effects of the beta particles (negatively charged electrons) emitted during decay.

The actual count rates and the current produced by the ionization events depend on a number of factors, including the nature and pressure of the gas, the dimensions of the detector, and the amount of voltage applied between the center wire and cylinder wall. Figure 4.3 illustrates the effect on pulse heights (amount of current produced) as a function of increasing counter voltage. In the *proportional region,* alpha and beta particles can be distinguished because the current produced is proportional to the energy of the ionizing radiation. In the *Geiger region,* however, this effect disappears as the size of the pulse produced by ionizing radiation becomes independent of its energy. In typical gas counting systems, proportional counting is employed for the detection of ^{14}C decay in the main sample counter, whereas anticoincidence counters operate either in the Geiger or proportional region (Section 4.3.1; Fig. 4.7).

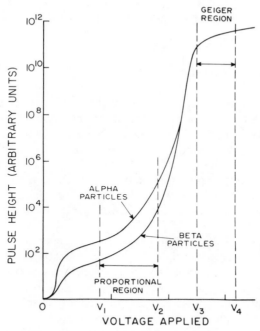

Figure 4.3 Proportional and Geiger counting regions of a typical gas detector. [After Friedlander *et al.* (1964:142).]

4.3 MEASUREMENT OF NATURAL RADIOCARBON

The challenge of accurately and precisely measuring natural ^{14}C concentrations in organic samples proceeds from the consequences of three facts. Most importantly, cosmic-ray-produced ^{14}C occurs in extremely low concentrations. The natural isotopic composition of modern carbon is about 98.9% ^{12}C, 1.1% ^{13}C, and 10^{-10}% ^{14}C. (The $^{13}C/^{12}C$ ratio varies somewhat depending on the source and geochemical history of the carbon compound.) Thus, one naturally produced ^{14}C atom exists for about every 10^{12} (1,000,000,000,000) ^{12}C atoms in *living* materials. This concentration decreases by a factor of two for every approximately 5700-year period following the death of the organism or withdrawal of the carbon-containing material from an active reservoir. Also, the beta particles emitted during decay are relatively weak. They have a maximum range of about 22 cm in air and 0.1 mm in aluminum foil (Raaen *et al.*, 1968). Finally, ^{14}C is a relatively long-lived nuclide. With a half-life of 5700 years, the mean or average life span of an individual ^{14}C atom is about 8000 years (Section 4.6). Extremely sensitive counting techniques had to be developed to obtain routine measurements of natural concentrations of ^{14}C.

4.3.1 General Considerations

In 1946, the initial critical experiment that demonstrated the difference in ^{14}C activity between fossil and modern carbon (Chapter 6) involved the use of a thermal diffusion column to artificially enrich the ^{14}C content of the samples so that a methane gas counter could be employed to make the measurements. The use of a thermal diffusion column on a routine basis would probably have been (and still is) impractical because of prohibitive costs. Methods to measure natural ^{14}C activities without recourse to enrichment had to be employed if ^{14}C was to be widely used as a practical dating method.

Three generations of detection systems have been applied to the problem of routinely measuring natural ^{14}C concentrations: (i) decay or beta counting of elemental carbon, (ii) decay counting using gas or liquid scintillation detectors, and (iii) direct counting using particle accelerators. All of the pioneering ^{14}C dating studies conducted by Libby and other early researchers used *solid carbon* in what was called a "screen-wall" type of counter. This technique was rapidly replaced by *gas counters* or *liquid scintillation* instruments. In gas systems, the sample material is converted to carbon dioxide (CO_2) or to a hydrocarbon gas synthesized from the CO_2. These gases were used directly as the counting gas in a counter. In contemporary liquid scintillation applications, samples are typically first converted to CO_2 and then synthesized in a series of relatively complex chemical steps to benzene (C_6H_6) to which is added a scintillator solution that produces light pulses in the presence of ionizing radiation. Extremely sensitive photomultiplier tubes detect and amplify the light pulses. Beginning in the 1980s, the third generation of ^{14}C technology began to include the routine use of accelerators employed as high-energy mass spectrometers to achieve direct or ion counting of ^{14}C.

In addition to these methods, several other approaches have been proposed for measuring low-level ^{14}C activities. They include the use of a bubble chamber (Aitken, 1974), thermoluminescence dosimetry (Ralph and Han, 1971:250; Winter, 1972), nuclear track plates (Jeffreys et al., 1972), and spectrometric analysis of samples enriched in ^{14}C using a finely tuned laser (Hall and Hedges, 1977:12; Hedges and Moore, 1978; Hedges et al., 1980). None of these approaches, however, have, to date, been developed to the point where routine measurements are possible on a practical basis.

All of the methods used to accomplish natural ^{14}C detection are concerned with the technical problems relating both to the *efficiency* and *background* of the systems employed to make the measurements. Efficiency relates to the number of events detected as opposed to the those that actually occur during a period of measurement. Solid carbon counting (Section 4.4.1 and 6.4) was relatively inefficient since it detected only about 5% of the decay events, whereas contemporary gas or liquid scintillation systems are typically 95–99% efficient. Counting backgrounds reflect environmental ionizing radiation effects that have to be distinguished from beta events derived from the decay of ^{14}C. Various sources contribute to background radiation, from various primary and secondary effects of cosmic rays to radioactive impurities in the materials from which the detectors are fabricated. Without some mechanism to reduce background activity, the count rate generated from environmental radiation would make it difficult to accurately measure the very low natural ^{14}C concentrations.

The general configuration of the instrumentation employed in gas proportional decay counting is outlined in Fig. 4.4. Typically, components include a shield assembly and detector assembly (comprising a sample detector and "guard" ring) as well as the counting electronics and high-voltage sources. In standard decay counting systems, the first line of defense against the effect of external radiation is shielding. Detector assemblies can be placed in basement locations of multistoried structures and surrounded by high-density materials such as iron, steel, lead, and/or mercury. In gas counting systems, various components of the background radiation can also be absorbed with special preparations, such as boric acid in paraffin, that absorb a significant percentage of the secondary neutron

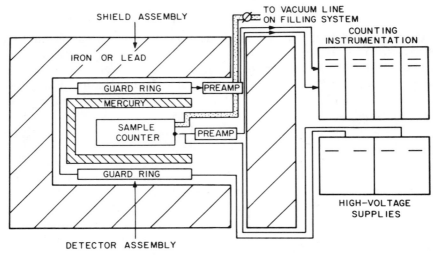

Figure 4.4 Components of a gas decay counting system.

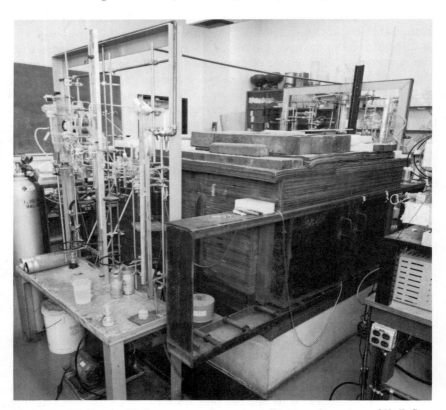

Figure 4.5 Shield assembly for a gas counting system. (Photograph courtesy of H. E. Suess and L. D. Ford, University of California, San Diego.)

Figure 4.6 Detector assembly including detector and concentric guard. (Photograph courtesy of H. E. Suess and L. D. Ford, University of California, San Diego.)

component of cosmic rays. In addition, detectors may be positioned underground to take advantage of the shielding properties of 5–10 m of soil. Figure 4.5 illustrates a shield assembly for a gas counting system. Figure 4.6 illustrates a single detector assembly including a detector and concentric guard separated by a cylindrical lead shroud. Above the detector assembly are situated paraffin/boric-acid blocks to reduce the neutron component of cosmic radiation.

Although physical shielding is crucial in low-level counting, historically the most important development that made the measurement of natural ^{14}C practical involved the utilization of an *anticoincidence* circuit linking together the detector containing the sample and a set of surrounding counters (a "guard ring") typically involving an assembly of Geiger–Müller

(GM) tubes or a concentric annular or rectangular guard, as illustrated in Fig. 4.7. The decay of ^{14}C or any other radioactive nuclide in the sample detector (as in event B in Fig. 4.7) will trigger only that detector. However, the central detector can also be triggered as a result of external radiation reaching the detector through the shield as in event A. To distinguish between these two possibilities, pulses from both surrounding counters and the central sample detector are compared electronically. If both are in coincidence (i.e., both occur within the time span of a few milliseconds or less), then the cause of the detector pulse is considered to be external to the central detector. However, if the sample detector generates a pulse *not* in coincidence (anticoincidence) with a pulse received from the surrounding ring within a given interval of time, then the event is considered to have occurred within the effective volume of the detector itself. Thus, only anticoincident pulses from the central counter are used to infer ^{14}C (and radon) activity.

4.3.2 Preparation of Samples

The initial step for all gas and most liquid scintillation systems is the conversion of the sample following pretreatment (Chapter 3) into CO_2 typically by combustion for noncarbonate materials or acidification for shell or other inorganic carbonates. The production of CO_2 takes place in a closed system, which had been first flushed with oxygen or some inert gas to remove atmospheric CO_2. This is necessary since air typically contains about 0.3% CO_2. This CO_2 contains contemporary ^{14}C that would mix with the CO_2 produced from the sample to yield an anomalous composite ^{14}C activity.

In most laboratories, combustion of organic samples is accomplished in some type of fused-quartz tube furnace arrangement in the presence of oxygen introduced under strictly controlled conditions. In some cases, samples are placed inside an inner combustion tube, which in turn is inserted into a larger outside enclosing tube. The purpose of the dual tube arrangement is to ensure that potentially explosive gases formed during the heating process can be safely and completely burned. This can be accomplished by initially directing an inert carrier gas (e.g., argon or nitrogen) into the central sample tube and oxygen into the enclosing tube. When heat is applied, the distilled combustive gases driven off the sample materials can be burned in a controlled manner as they exit into the oxygen atmosphere in the outer tube.

In several ^{14}C facilities, a high-pressure combustion chamber, or "bomb", is employed as the preferred method of producing CO_2 from samples (Burleigh, 1974). This typically consists of a stainless steel cylinder

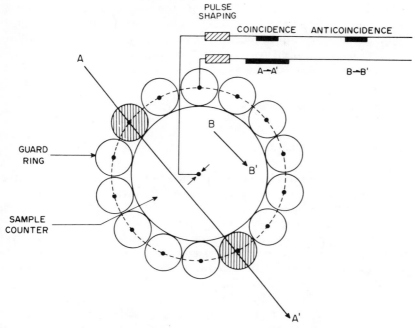

Figure 4.7   Representation of the anticoincidence principle used in low-level decay counting.

designed to withstand relatively high pressures (1500–2000 lb/in^2). After a sample is placed within the combustion unit, the bomb is sealed and flushed to remove atmospheric gases. It is then pressurized with pure oxygen and an extremely rapid combustion event is initiated by passing an electrical current through a high-resistance coil. The combustion bomb method is most useful when relatively large sample sizes are available. For one laboratory, the optimum sample size is about 12 g of carbon, although samples as small as 1 g can be accommodated (Burleigh, 1974:301). For samples sizes of less than 1 g, combustion in a tube furnace is preferable.

For carbonate samples such as shell, CO_2 is released by evolution with an acid, usually hydrochloric (HCl). The typical experimental arrangement employs a carrier gas that is used first to flush the flask containing the sample of atmospheric gases and then to assist in the transport and collection of the CO_2 evolved from the sample. In some laboratories, a "wet combustion" is used for certain types of samples, such as muds exhibiting low organic carbon content (Albero and Angidini, 1985). In such cases, the sample is digested in a glass vessel with strong oxidizers such as sulfuric acid (H_2SO_4) and sodium dichromate ($Na_2Cr_2O_7$). By whatever process, the CO_2 obtained from a sample is transported through a purification train before being condensed in traps on a vacuum line using liquid nitrogen or liquid air. Purification of the CO_2 using a variety of oxidation and reduction steps is required to remove such impurities as water vapor, products of incomplete combustion, and oxides of nitrogen and sulfur.

4.4 DECAY COUNTING SYSTEMS

4.4.1 Solid Carbon Counting

The pioneering research that established the ^{14}C method was carried out by Libby and his co-workers, James R. Arnold and Ernest C. Anderson, using a Geiger counter and employing a "screen wall" design. Carbon dioxide obtained from sample combustion/acidification was converted by reduction on magnesium to elemental or solid carbon, "carbon black." This elemental carbon was used to coat a portion of the inside surface of a sleeve that was inserted into the counter. Generally, about 10–12 g of carbon were required for each sample and the maximum dating range that could be obtained was about 25,000 years with typical counting errors of 200–300 years for samples up to about 5000 years old. The dates produced at the University of Chicago were first published in the journal *Science* and subsequently in two editions (1952 and 1956) of Libby's *Radiocarbon Dating*. All of these values were obtained using the solid carbon technique. Many of the laboratories that attempted to duplicate the original solid carbon method experienced different types of difficulties at varying levels of severity. A major problem was the contamination of sample preparations with fission products from the detonation of thermonuclear devices in the atmosphere. The Chicago laboratory completed its work in 1956 when Libby left to take a position with the United States Atomic Energy Commission. By this date, almost all laboratories which had begun with a solid carbon approach or were contemplating ^{14}C research had decided to employ some type of gas counting system or were investigating liquid scintillation counting. Thus, solid carbon decay counting has not been used for several decades. A brief discussion of its technical aspects is outlined in Chapter 6 in the context of a discussion of the historical development of ^{14}C dating.

4.4.2 Gas Counting

By the mid-1950s, difficulties with the solid carbon approach combined with greater experience with gas counting for other nuclear research applications led to the use of gas counting for natural ^{14}C measurements. Three counting gases are widely employed: carbon dioxide (CO_2), methane (CH_4), and acetylene (C_2H_2). One laboratory also uses ethane (C_2H_6). Each approach has specific advantages and disadvantages. The choice of systems is largely based on the training and experience of different investigators.

As noted, the immediate stimulus to replace solid carbon counting was difficulties encountered by various researchers as they attempted to duplicate the original analytical procedures during a period when fallout products contaminated sample preparations. Over the longer term, however, the principal advantage of gas (and liquid scintillation) counting was the significant increase in practical operational efficiency over that of solid carbon counting. There was also an increase in maximum dating range and statistical precision, as well as a reduction in the amount of sample material required for analysis. In general, the maximum dating limit for gas systems range from 30,000 to 60,000 years depending on backgrounds exhibited by counters, counter sizes, counting pressures employed, and typical counting times. With isotopic enrichment, maximum finite ages of up to 75,000 years have been reported (Grootes *et al.*, 1975; Stuiver *et al.*, 1978; Erlenkeuser, 1979). Sample size requirements with gas systems with operational limits up to about 40,000 years range from about 1 to 5 g of carbon. Special high-precision instruments designed primarily for geophysical studies can achieve one-sigma, or 1 σ, (i.e., one standard deviation) statistical ranges of 15–20 years but require 7–20 g of carbon and, in some cases, special experimental configurations including underground locations for detectors to reduce backgrounds. Laboratories employing gas systems using from 0.5 to 5 g of carbon after pretreatment report 40–100 years 1 σ counting variance for samples up to 5000 years old.

As we have noted, the initial step for all gas systems is the conversion of the sample into CO_2 by combustion or acidification. Ralph (1971:31–36) has provided a convenient summary of the basic steps involved. Figure 4.8 illustrates a vacuum line used for the combustion and purification of CO_2 sample gases. For laboratories employing CO_2, its use as a counting gas requires that it be of very high purity. Extremely small traces of electronegative impurities such as oxygen, the halogens (e.g., chlorine, bromine, fluorine), and oxides of nitrogen and sulfur significantly disturb the counting characteristics of CO_2 for beta detection (de Vries and Barendsen, 1953; Rafter, 1955a; Fergusson, 1955; Olsson, 1958). Its use for ^{14}C counting requires that stringent chemical treatment remove contaminants to below about 1–10 parts per million (ppm).

Laboratories employing CO_2 have developed a variety of techniques designed to achieve the required gas purity. One of the earliest approaches

Figure 4.8 Diagram of conventional vacuum line for preparation and purification of gases for ^{14}C analysis.

purified CO_2 by reacting it with calcium oxide (CaO) at about 700°C. Some of the CaO reacts with the CO_2 to form calcium carbonate ($CaCO_3$). Some of the impurities also react with the CaO. The tube containing the CaO and $CaCO_3$ is then allowed to cool and the remaining impurities, which do not react with the CaO, are pumped off. The temperature is then raised again over 700°C and the CO_2 is regenerated and collected (Rafter, 1955a). Many laboratories have found that this approach is generally not needed if the sample is passed a number of times through a thoroughly reduced copper furnace maintained at 400°C.

A drawback of the CaO process is that it frequently adds radon to the sample. Since radon has a half-life of about 3.8 days, storing a CO_2 sample for an appropriate period of time usually allows the radon to decay below the limit of detection. Some laboratories have devised methods of removing the radon by taking advantage of the fact that its freezing point is slightly different from that of CO_2 (de Vries, 1957). By lowering the temperature to precisely the freezing point of CO_2, the radon can be pumped off. Unfortunately, the process is not 100% efficient and slight losses in CO_2 tend to occur. Several laboratories employ instrumental methods that permit a small amount of radon to be measured during counting by subtracting the contribution of the radon daughter betas from the total beta count rate (Fergusson, 1955). However, the majority of laboratories solve this problem by the storage of their gas samples for a period from 3–8 weeks (Nydal, 1983b). Interestingly enough, one laboratory reported variations in count rates traced to the presence of natural radon emanating from the soil surrounding a basement facility (Freundlich, 1973).

Some early researchers, aware of potentially severe purity problems with the use of CO_2 as a counting gas, examined other counting approaches. The University of Michigan laboratory, which focused its attention on archaeological samples throughout most of the period of its operation, employed a unique carbon dioxide–carbon disulfide (CO_2–CS_2) gas counting system (Crane, 1956). An advantage of such a system was the ability to operate in the Geiger region. Because of the relatively large pulses, interference from noise and disturbances in electronic circuits was minimized, since little amplification of the pulses was required. The use of carbon disulfide, in very small amounts, allowed the use of relatively impure CO_2 to be employed and yet relatively stable and reproducible count rates were reported.

Other investigators developed methods to convert the CO_2 obtained from sample preparations to a hydrocarbon gas that exhibited more acceptable counting characteristics. Methane (CH_4) is one of these hydrocarbons that is much less sensitive to electronegative impurities. However, the preparation of methane involves chemical conversion steps that, if not carefully carried through to completion, carry the potential of carbon isotope fractionation in the gas preparation. Gas handling techniques are also more difficult than with CO_2. Production of CO_2 by combustion or acidification is conducted as described previously with particular care to remove any traces of water vapor. Dry CO_2 is reacted with hydrogen in the presence of a heated ruthenium catalyst. One of the products of this reaction is methane along with water, which is continuously trapped out during the chemical reaction (Fairhall et al., 1961; Long, 1965). Some problems in obtaining hydrogen that is free of tritium, the radioactive isotope of hydrogen, were initially reported.

Acetylene (C_2H_2) is also used by several laboratories as a counting gas for ^{14}C work. Like methane, the counting characteristics of acetylene are less sensitive to the effects of impurities than are those of CO_2. Acetylene has an added benefit in that each molecule of acetylene contains two carbon atoms. Thus, there is a doubling of ^{14}C activity for the same counter volume. On the other hand, the preparation of acetylene involves a number of steps that, like those for methane, must be carried through to completion to avoid fractionation. The preparation of acetylene for ^{14}C counting from CO_2 from combustion/acidification as described by Suess (1954a,b) involves the absorption of CO_2 in ammonium hydroxide (NH_4OH) to form ammonium carbonate ($(NH_4)_2CO_3$). Strontium chloride ($SrCl_2$) is used to precipitate out strontium carbonate ($SrCO_3$). The strontium carbonate is washed, filtered, dried, and then mixed with magnesium powder. This mixture is then placed into a stainless steel tube, evacuated, and heated. The reaction that occurs produces strontium carbide (SrC_2). Acetylene is produced by the addition of water inside the evacuated system. The acetylene is collected in a liquid nitrogen trap. After purification over charcoal, it is usually stored for two to three weeks to allow any radon to decay.

Ethane (C_2H_6) can be produced by the reaction of acetylene with hydrogen in the presence of a palladium catalyst. Depending on the purity of the acetylene, some ethylene and butane are formed, which affects counting characteristics. However, purification can be accomplished by freezing out the ethane in degased charcoal. For one laboratory, ethane was considered the most satisfactory counting gas since it required the lowest working voltage and gave the longest and flattest counting plateaus (Geyh and Schneekloth, 1964; Geyh, 1965).

4.4.3 Liquid Scintillation Counting

Parallel to the development of gas counting in the early 1950s, several investigators examined the potential of employing liquid scintillation as a means of measuring natural ^{14}C (e.g., Hayes et al., 1953; Audric and Long, 1954). Although investigated at Chicago by Arnold (1954) and used in the

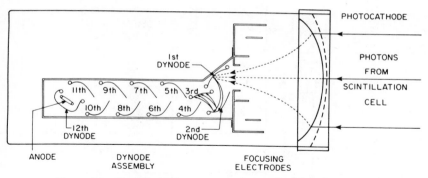

Figure 4.9 Simplified schematic of 12-stage photomultiplier tube.

late 1950s at the ^{14}C facility at Trinity College, Dublin, Ireland (Watts, 1960), liquid scintillation counting did not begin to be widely employed for natural ^{14}C measurements until well into the 1960s. This technique takes advantage of the fact that, in the presence of ionizing radiation, certain types of organic compounds (scintillators) emit short bursts of light energy (photons). This energy can be converted into an electrical current in a photomultiplier tube. The principle behind this device is the photoelectric effect, in which electrons are dislodged from a metallic surface during exposure to light. Photomultiplier tubes were designed to permit the measurement of very low-intensity light energy by multiplying the effect of a single photon (Horrocks, 1976).

The operation of a photomultiplier tube is illustrated in Fig. 4.9. In this case, twelve electrodes (dynodes) are placed in a lineal arrangement with each dynode at a higher potential voltage than the preceding one. When photons impinge on the photocathode, electrons are produced. In our example, an electrostatic field focuses these electrons onto the first dynode. Additional electrons are produced at each of the dynodes and then some are collected at the anode. The current produced in this device is then conducted to appropriate circuitry and further amplified.

Initially, liquid scintillation ^{14}C measurements typically involved the use of a scintillator known as PPO (2,5-diphenyloxazole) dissolved in toluene as the solvent, while ethanol, methanol, or methyl borate are examples of solvents used for the sample (Hayes *et al.*, 1953; Arnold, 1954; Pringle *et al.*, 1957; McAuley and Watts, 1961; cf. Noakes, 1976:190). Other approaches included dissolving acetylene in toluene (Audric and Long, 1954) and using CO_2 in a liquid form dissolved in toluene at very low temperatures and then maintained in the sample cell at relatively high pressures and $-20°C$ (Barendsen, 1957). Figure 4.10 provides an example of the counting system employed in early scintillation counting (Arnold,

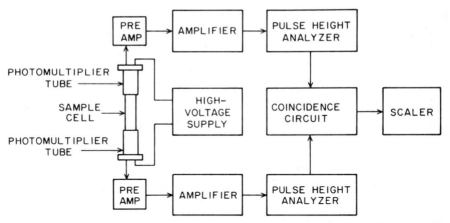

Figure 4.10   Basic elements of liquid scintillation counting instrumentation for ^{14}C analysis.
[After Arnold (1954:156).]

1954:156). The use of two photomultiplier tubes linked together into a coincidence circuit allowed spurious pulses from the tubes to be identified. In early systems, electronic noise in the photomultiplier tubes, "dark current noise," was reduced by lowering the ambient temperature by refrigerating the tube assembly.

Since these initial studies, liquid scintillation ^{14}C counting has undergone continuous development in both the techniques of synthesis of the sample solvent and the optimization in the design of liquid scintillation instrumentation (Polach, 1974; Noakes, 1976). These refinements have been greatly facilitated by the rapid development of commercial liquid scintillation spectrometers for use in biological and biomedical research (Rapkin, 1969). Modifications in the commercially available systems were typically made to optimize the features that were important in measuring natural ^{14}C concentrations. These modifications, many of which were designated to lower backgrounds, included, for example, the use of increased shielding for the detector photomultiplier assembly (Noakes *et al.,* 1974; Hartley and Church, 1974), improvements in the design and materials used in the fabrication of the vials used to hold the sample solvent and scintillator solute solutions (Calf and Polach, 1973; Hass, 1979), and optimization of the electronic components of the spectrometer, including the use of an anticoincidence guard system (Gupta and Polach, 1985:50–64).

Since the early 1960s, benzene (C_6H_6) has most often been used as the sample solvent in low-level ^{14}C liquid scintillation counting. An advantage of benzene is that it contains over 90% carbon and has excellent scintillation properties. Benzene synthesis generally involves the production of

CO_2 from a sample, the conversion of CO_2 to acetylene (C_2H_2), and the conversion of acetylene to benzene. A widely employed technique involves the synthesis of lithium carbide (Li_2C_2) from CO_2 and its hydrolysis to acetylene (Barker, 1953). The purity of the benzene produced in the relatively complex chemical steps employed must be carefully monitored since trace impurities at the parts-per-million level can significantly affect counting efficiency (Fraser *et al.*, 1974). A typical measurement involved the use of benzene to which was added PPO and sometimes POPOP. The later solute was used to shift the spectrum of the emitted light to a region most sensitive to earlier types of photomultipliers (Tamers, 1965:54–55). Newer scintillators have been developed to enhance efficiency and improve the performance of the photomultipliers (Polach *et al.*, 1983).

An important advantage of liquid scintillation for natural or low-level ^{14}C analysis, as compared with gas systems, is a reflection of the much higher sample density. This permits reduction in counting chamber dimensions and thus a reduction in relative background rates. In addition, the nature of background pulses in liquid scintillation systems permits the use of pulse height analysis to distinguish background from ^{14}C pulses. A further practical advantage is the ability to rapidly cycle vials containing sample, background, and contemporary standard solutions in and out of the counting chamber. This greatly facilitates the analysis of counting data as well as the quick identification of any instrumental malfunctions (Polach *et al.*, 1984). Originally, liquid scintillation instruments required sample sizes somewhat in excess of that normally required in gas counting. However, recent developments have permitted significant reductions so that some liquid scintillation systems can now accommodate sample sizes below 1 g of carbon on a routine basis, although at least one group reports that, in reducing sample sizes to that level ". . . more insidious unquantifiable errors, apparently indigenous in the LS [liquid scintillation] method, tend to become magnified in effect as the sample size diminishes" (Otlet and Evans, 1983:216).

For many investigators, liquid scintillation is considered the method of choice for decay counting ^{14}C analysis. However, some concerns have been expressed about the relatively complex chemistry and exacting attention to detail required to process samples and the relatively large number of analytical and instrumental parameters that must be closely monitored to maintain stability and obtain precise natural ^{14}C measurements (cf. Tamers, 1965; Noakes *et al.*, 1965; Pearson *et al.*, 1977; Otlet and Warchal, 1978; Pearson, 1979:4–5). The success of most laboratories using scintillation counting in maintaining these standards, however, may be reflected in the fact that the majority of ^{14}C dating facilities installed over the last decade employ liquid scintillation instrumentation even though the majority of laboratories still employ gas counting systems (Mook, 1983a; cf. Browman, 1981:245).

4.5 DIRECT COUNTING SYSTEMS

In conventional decay counting systems, a very small fraction of the ^{14}C atoms present are actually measured. There are, for example, approximately 5.9×10^{10} atoms of ^{14}C in 1 gram of modern "prebomb" carbon. However, on the average, over a one-minute period, less than 14 of them will decay and be available for detection. In large part, it was this consideration that gave impetus to efforts to develop direct counting for ^{14}C. It was recognized that significantly higher efficiencies of atom-by-atom detection such as that employed in mass spectrometers would allow the use of much smaller samples. At the same time, it would provide a potential method of extending the maximum ^{14}C age range beyond that generally possible with decay-counting techniques (Muller, 1977).

As early as 1970, Oeschger and his co-workers noted the great increase in sensitivity that could be obtained with the use of mass spectrometry combined with isotopic enrichment (Oeschger *et al.,* 1970:487–488). Attempts to make direct counting measurements using a conventional mass spectrometer could not be developed for practical operation because ^{14}C was masked by stable ions with similar charge-to-mass ratio in the mass spectrum (Anbar, 1978; cf. Wilson, 1979). In the late 1970s, successful direct counting of ^{14}C was accomplished by accelerating sample atoms in the form of ions to higher energies in particle accelerators. Initially, the term high-energy mass spectrometry (HEMS) was employed to describe this approach. Now, more commonly, the phrase accelerator mass spectrometry (AMS) is used. For energies on the order of 0.5 MeV per nucleon, ions of the same mass and energy but differing nuclear charge (e.g., ^{14}C and ^{14}N) can be distinguished by measuring the total energy of each ion and its rate of energy loss. In electrostatic accelerators, another important advantage of going to highly charged states (3 + or higher) is that all molecules are destroyed in the stripping process. Unfortunately, the extremely sensitive detectors employed in such nuclear physics experiments cannot function with beams in excess of a few thousand particles per second. Because of this, elimination of as many unwanted particles as possible must be accomplished before or during acceleration by magnetic or electrostatic means (Litherland, 1980, 1984; Kutschera, 1983).

The operation of an AMS system to measure natural ^{14}C concentrations involves four basic steps: (i) production of ions from a sample in an ion source, (ii) acceleration of the ionized particles, (iii) separation of the ^{14}C from other isotopes and molecules, and (iv) counting of individual ^{14}C ions. Two types of particle instruments have been employed in direct ^{14}C measurements: cyclotrons and electrostatic accelerators. Each system employs different nuclide separation and identification strategies. However, both share a common feature in that a significant amount of attention has been focused on the design of the ion source since its performance plays a crucial role in efficient operation of an AMS system. Characteristics of an ion source that must be considered include (i) beam current (the number of ions produced per unit time), which must be relatively high to minimize run times, (ii) efficiency of the source, (iii) stability, (iv) ability to avoid isotopic fractionation, and (v) ability to avoid memory effects (Hedges, 1981).

In cyclotrons, high energies are imparted to particles by accelerating ions through two semi-circular high-voltage electrodes within a magnetic field (Stephenson *et al.*, 1979). An alternating accelerating voltage is applied between the electrodes and repetitive acceleration continues until the particles reach an energy sufficient to move to the edge of the magnet where a beam can be extracted. An important characteristic of the cyclotron is that magnetic separation of different ions takes place simultaneously with acceleration, i.e., the process of acceleration in the cyclotron itself acts as a charge-to-mass ratio filter. When the cyclotron frequency is tuned to accelerate ^{14}C, the only other ion present is ^{14}N. Other ions with the wrong charge-to-mass ratio quickly drop out of phase and are lost from the beam. As illustrated in Fig. 4.11, to remove the ^{14}N from the beam, a "range separation" method can be employed. This method takes advantage of the fact that the distance traveled by ^{14}N in a solid or gas is about 30% less than that of ^{14}C. A gas cell or metal foil placed in the beam line is used to discriminate against ^{14}N. The ^{14}C ions are detected by an ionization chamber and a solid-state device to obtain the total energy and energy loss of the particles (Mast and Muller, 1980).

Initial experiments performed to date with cyclotrons to measure ^{14}C have used positive ions obtained from a CO_2 gas ion source. This is an advantage in that gas samples can be prepared easily in a closed system free of possible contamination. A negative ion C^- source has also been discussed. Beam currents using gas ion sources are characteristically lower than sources using solid samples and backgrounds are higher, but studies to overcome this problem have been initiated (Middleton, 1984). Studies are also underway to develop a lower-energy (40-keV) negative ion cyclotron for direct detection of ^{14}C (Welch *et al.*, 1984).

In an electrostatic accelerator, the voltage required to accelerate the ^{14}C ions to high energies is provided either by a moving belt or chain (Van de Graaff type) or by a solid-state voltage multiplier (Cockcroft–Walton

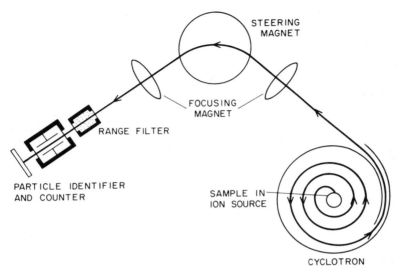

Figure 4.11   Principle of use of cyclotron as a mass spectrometer. Schematic not to scale. [From Mast and Muller (1980).]

type). Negative ions are accelerated toward a positively charged terminal where electrons are removed by passage through a gas "stripper" to create positively charged ions, which are then accelerated away from the terminal. In the stripping process, no molecules survive at charge states $3+$ or higher. The total energy of the ion is a result of energy acquired during acceleration both toward and away from the terminal. Since the acceleration occurs in two stages such an instrument is referred to as a *tandem accelerator*. The use of tandem accelerators for accelerator mass spectrometry is known as tandem accelerator mass spectrometry (TAMS). To illustrate the characteristics of this type of AMS system, Fig. 4.12 represents the beam-line arrangement of a tandem accelerator operating at the University of Arizona (Tucson). Figure 4.13 provides a photographic view of the accelerator components oriented as in Fig. 4.12 with the beam line proceeding from right to left.

One of the important advantages of a tandem accelerator is that ^{14}N apparently does not form negative ions that live long enough to pass through the accelerator to the detector (Bennett *et al.*, 1977; Purser *et al.*, 1982). Thus an important source of background is essentially eliminated. Another important characteristic is the fact that it is a relatively simple operation to sequence the acceleration of the various isotopes of carbon through the system in a reproducible manner. This permits measurements to be made of $^{14}C/^{13}C$ and $^{13}C/^{12}C$ ratios in times that are short compared to the timing of changes in system transmission characteristics.

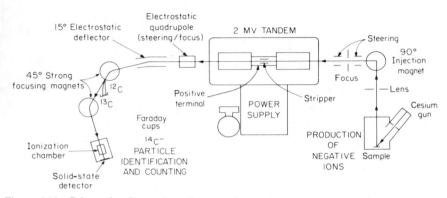

Figure 4.12 Schematic of a tandem electrostatic accelerator system used in accelerator mass spectrometry for direct ^{14}C measurement at the University of Arizona. Schematic not to scale. [From Taylor *et al.* (1984a).]

Currently, all routine AMS ^{14}C measurements employ solid samples in a variety of physical forms as the target in an ion-sputter source. Target materials have included various types of graphitic preparations, carbides, metal mixes, and metal solutions (Polach, 1984). Solids have been preferred as target materials because of the higher beam currents and memory effects with gas sources. However, preparation of solid targets has introduced problems of sample contamination that probably could be more easily controlled if CO_2 could be employed. Research is underway to develop a CO_2 ion source that will yield beam currents comparable to solid sources and avoid memory effects (Heinemeier and Anderson, 1983; Middleton, 1984).

Figure 4.13 Accelerator mass spectrometer for direct counting of ^{14}C. (Photograph courtesy of T. Jull.)

The earliest major publications discussing the use of accelerators to measure ^{14}C directly appeared in 1977 (Muller, 1977; Nelson *et al.*, 1977; Bennett *et al.*, 1977). The first published ^{14}C determination *on an archaeology-related sample* by direct counting was obtained on the 88-in. (224-cm) cyclotron at the Lawrence Berkeley Laboratory (Muller *et al.*, 1978). A value of 5080 ± 60 ^{14}C years B.P. (UCLA-2118D) had previously been obtained by conventional decay counting on a sample of charcoal from the site of Pikimachay Cave in Peru (R. Berger, personal communication). The initial Berkeley AMS ^{14}C age estimate, obtained in a blind experiment, i.e., without the knowledge of the UCLA value, was 5900 ± 800 years B.P.

Since 1978, there has been a dramatic expansion in AMS research as applied to the measurement of a number of radioactive isotopes including ^{14}C. A number of review papers have appeared that set forth the history of the development of AMS technology, including that applied to ^{14}C analysis (e.g., Muller, 1979; Bennett *et al.*, 1978; Bennett, 1979; Mast and Muller, 1980; Gove, 1981; Litherland, 1980, 1984; Kutschera, 1983). Since 1979, developments in AMS ^{14}C studies have been extensively discussed at the international radiocarbon conferences (Chapter 1, Table 1.1, Part A), and at symposia called to review AMS developments (Chapter 1, Table 1.1, Part B). Discussions concerned with the utilization of AMS ^{14}C technology for the analysis of archaeological samples include those of Haynes (1978), Berger (1979), Pavlish and Banning (1980), Hedges (1981), Taylor *et al.* (1984a, 1984b), Gillespie *et al.* (1984a), Donahue *et al.* (1983, 1984), and Hedges and Gowlett (1986).

The majority of the discussions that have dealt with AMS ^{14}C analysis to date have understandably focused attention on issues and problems in the development of AMS instrumentation, specifically, the design of the ion source, as well as the optimization of accelerator and detector hardware. Equally important, however, are those issues that concern the appropriate methods for the routine handling of milligram-size samples (Hedges, 1983). The collection and chemical pretreatment of microsamples intended for AMS analysis must operate under constraints much more stringent than those routinely employed in most decay counting laboratories. In some cases "clean room" environments may increasingly be required for AMS sample processing, particularly if the potential of the AMS method to extend the ^{14}C time scale is to be utilized. Current maximum ages that can be measured on AMS systems range between 40,000 and 60,000 years. These values are set by the background levels presently being experienced in the accelerators. A significant component of these backgrounds appears to be small amounts of contamination not eliminated during pretreatment or introduced during the processing of the samples. As previously noted, current procedures require the conversion of samples to some solid carbon form that can be efficiently employed in a standard ion source. Close attention to detail in sample handling procedures, along with the use of an efficient, memory-free gas ion source (e.g., Middleton, 1984) may reasonably be expected to result in significant reductions in background levels.

Table 4.1 summarizes some important characteristics of the three generations of instrumentation employed in ^{14}C studies since the inauguration of routine analysis. Decay counting of elemental or solid carbon was replaced in the mid- to late-1950s by decay counting employing gas or liquid scintillation systems. In the succeeding thirty years, four types of decay counting emerged: (i) *standard* decay counting with sample sizes ranging up to about 5 g of carbon and with a maximum range of up to about 40,000 years, (ii) *high precision,* (iii) *extended range* (cf. Stuiver, 1982; Stuiver *et al.,* 1979) instrumentation requiring relatively large amounts of sample material (in the range of 10–20 g of carbon), and (iv) *microsample* analysis with micro- (5–10 ml) or mini- (>10 ml) detectors (cf. Otlet *et al.,* 1983) that can accommodate as little as 5–10 mg of carbon (cf. Harbottle *et al.,* 1979; Sayre *et al.,* 1981; Otlet and Evans, 1983). The third generation of instrumentation employs direct counting by AMS techniques utilizing milligram amounts of carbon.

4.6 RADIOCARBON AGE CALCULATIONS

Regardless of the type of instrumentation employed, a ^{14}C age calculation requires that four values be available for computation, three of which are unique to each counting system. These unique values include the background count rate, the count rate of the contemporary standard, and the count rate of the unknown age sample—all in the same detector operating under a common set of experimental conditions. The decay constant of ^{14}C, which is directly related to the half-life figure, is the fourth value required.

Each detector has its own characteristic background count rate. This value is determined by employing a carbon-containing sample for which it can safely be assumed that measurable ^{14}C activity is absent. In most cases, this involves the use of a geologically old material such as a fossil fuel derivative (e.g., coal, CO_2 from gas wells, or other petroleum sources) or carbonates from a source of known age (e.g., Tertiary or Mesozoic limestone or other geologically ancient carbonate deposit). The count rates

TABLE 4.1
Radiocarbon Dating Instrumentation: Generation, Mode, and Type

Generation	Mode	Type	Typical sample size[a] (mg carbon)	Typical precision[b] (years)	Typical counting period (hr)	Approximate maximum range (years × 10³)
I	decay	solid[c]	2000–5000	±200–500	48–120	25
II Standard	decay	gas/liquid scintillation	250–5000	±40–150	24–72	30–40
High precision	decay	gas/liquid scintillation	10000–20000	±20–40	72–168	—
Extended range	decay	gas/liquid scintillation	10000–20000	±200–500[d]	72–168	70–75
Micro/mini	decay	gas	10–250	±100–200	960–2760	—
III	direct	solid[e]	2–5	±80–400	1–2	40 (100)[f]

[a]After pretreatment.
[b]One sigma counting error.
[c]Obsolete.
[d]For 20,000–40,000 year range.
[e]CO_2 gas source under development (Middleton, 1984).
[f]Potential to extend ^{14}C time frame to approximately 100,000 years (see text).

96

measured when such samples are introduced into a decay counting system detector derive from ionizing radiation produced by radioactivity from isotopes of uranium, thorium, or potassium contained in trace amounts in the walls of the counting chamber or from components of the cosmic-ray flux that have not been detected by the guard ring. The necessity that samples used for purposes of measuring the background be rigorously pretreated to exclude contamination is indicated by reports of measurable ^{14}C activity in samples such as coal and graphite.

One of the assumptions of the ^{14}C method is that all living biological materials exhibit the same ^{14}C activity. If this had been the case, each laboratory could have used some living biological product for its contemporary standard. The fossil fuel (or Suess) effect and later the detonation of thermonuclear explosives in the atmosphere (atomic bomb effect) created a situation where the activity of living biological materials has been altered by recent human activity (Section 2.5). Careful selection and pretreatment of nineteenth century wood rings and the extrapolation of the count rate of such samples to A.D. 1950, i.e., 0 B.P. (e.g., Bannister and Damon, 1972) have been used to circumvent this problem. However, many ^{14}C laboratories take advantage of contemporary standards distributed by the United States National Bureau of Standards (NBS). In 1956, at the request of James R. Arnold, a 1000-lb lot of oxalic acid (HOOC-COOH) was prepared from a 1955 crop of French sugar beets (Cavallo and Mann, 1980; J. R. Arnold, personal communication). Ninety-five percent of the count rate of this material, normalized to -19 per mil $\delta^{13}C$ (see Section 5.3.2), was used to define zero ^{14}C age or "modern" (Flint and Deevey, 1961; Olsson, 1970b, cf. Craig, 1961). This provided a reference as to what the contemporary or zero age of the biosphere would have been in the absence of the Suess and atomic bomb effects. A number of secondary standards have been related experimentally to the original NBS standard (e.g., Polach, 1979; Gupta and Polach, 1985:14). By the late 1970s, the supply of the original batch of oxalic acid was depleted and a second or "new" NBS contemporary ^{14}C standard was prepared. It was determined that 73.68% of the "new" NBS oxalic acid ^{14}C standard count rate was equilivant to 95 percent of the old when both were normalized to a $\delta^{13}C$ value of -19 per mil (Stuiver, 1980). It was later suggested that -25 per mil be used as the $\delta^{13}C$ value to which the new standard would be normalized (Mann, 1983).

An aspect of the introduction of the original NBS standard was that its use implied that A.D. 1950 constitutes 0 B.P. in ^{14}C age computations. The choice of A.D. 1950 was, to a degree, arbitrary; it was adopted to honor the publication of the first ^{14}C dates in December 1949 (Arnold and Libby, 1949). The agreement to use A.D. 1950 as the calendar zero reference year for ^{14}C dating redefined the meaning of B.P. from "Before Present" to "Before Physics" (Flint and Deevey, 1962). With time, of course, the ^{14}C activity of any standard including the NBS standard will decrease. However, as pointed out by Stuiver and Polach (1977:356), standards and samples will lose their ^{14}C at the same rate; the *ratio* between standard and sample will remain constant. Thus, in ^{14}C age calculations, there is no need to correct for changes in specific ^{14}C activity in either the primary NBS standards or in any standard related to them.

The mathematical computation of a ^{14}C age value proceeds from well-known principles of radioactive decay. The fundamental relationship is the familiar exponential decay equation, which, in several forms, can be found in most undergraduate physics texts or standard reference works (e.g., Friedlander *et al.* 1964:6–8).

$$A = A_0 e^{-\lambda t} \tag{4.1}$$

This expression allows the calculation of the activity A of a sample after a period of time t has elapsed when the original activity A_0 of that sample and the *decay constant* λ of that particular radioactive nuclide is known. The expression $e^{-\lambda t}$ describes the exponential rate of decrease observed in the decay of all radioactive nuclides. The symbol e in this equation is the base of the natural logarithm. A discussion of how it is derived and defined can be found in any introductory calculus text (e.g., Swokowski, 1975:289–326). It is approximately equal to 2.71828. A decay constant is the fraction of the number of atoms transformed by radioactive decay per some time unit. It is related to the half-life by the relationship:

$$\lambda = (\ln 2)/t_{1/2} \tag{4.2}$$

(The natural logarithm to the base e of 2 (ln 2) is 0.693). A third value seen in ^{14}C age calculations is the mean-life T. The mean-life expresses the "average life" of a given radioactive nuclide. It can be calculated by taking the reciprocal of the decay constant:

$$T = 1/\lambda \tag{4.3}$$

The relationship between the mean-life, decay constant, and half-life can be summarized in Eq. (4.4):

$$T = 1/\lambda = t_{1/2}/0.693 \tag{4.4}$$

The relationship between the mean-life and half-life is illustrated in Fig. 4.14. Table 4.2 summarizes the values of the mean-life and decay constant for the 5568 (5570) and 5730 half-life values of ^{14}C.

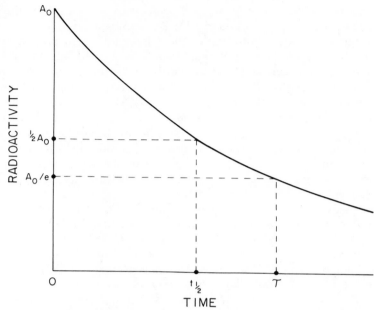

Figure 4.14 Relationship between the half-life ($t_{1/2}$) and mean-life (T) value of a radioactive isotope. [After Libby (1970a:10).]

By solving for t in Eq. (4.1) and substituting T in place of λ, we have a simple formula that can be used to calculate a ^{14}C age:

$$t \text{ (years)} = T \ln(A_0/A) \tag{4.5}$$

If we use the mean-life associated with the Libby half-life, the expression becomes

$$t \text{ (years)} = 8033 \ln(A_0/A) \tag{4.6}$$

TABLE 4.2
Values of Radiocarbon Constants

Constant	Value (years)	
	("Libby")	("Cambridge")
Half-life ($t_{1/2}$)	5568 (5570)	5730
Mean-life (T)	8033	8267
Decay constant (λ)	$\dfrac{1}{8033}$ or 1.2449×10^{-4}	$\dfrac{1}{8267}$ or 1.2096×10^{-4}

In expressions (4.5) and (4.6), A_0 is the accepted percentage of the net activity of a contemporary ^{14}C standard, e.g., NBS oxalic acid or a standard with a known relationship to one of the NBS oxalic acid preparations. For example, with the "old" NBS standard, 95% of its net activity defines ^{14}C activity at "time zero." These standards can be represented as "100% of modern" (modern = 0 B.P. = A.D. 1950) with every -1.0% deviation representing about an 80-year variation, i.e., 1% of the *mean-life* of ^{14}C, which, as we have noted, is about 8000 years. The net rate referred to is the actual or gross count rate minus the value of the background rate in the same counter under the same experimental conditions. The net count rate of the sample A is the gross rate minus the background rate in the same detector under the same experimental conditions as that at which the A_0 and background rate were measured.

The expression in Eq. (4.6) involves only the ^{14}C activity of a sample and standard to derive a ^{14}C age estimate. As discussed in Section 5.3.2, natural variations in stable carbon isotope ratios ($^{13}C/^{12}C$) must also be taken into account in the calculation of conventional ^{14}C age estimates (Stuiver and Polach, 1977:360). For the measurement of $^{13}C/^{12}C$ ratios, Harmon Craig (1953) employed a calcium carbonate belemnite (the fossil mollusc *Belemnitella americana*) from the Cretaneous age Peedee Formation of South Carolina as a standard (Craig, 1953, 1957). The Peedee belemnite (PDB), or Chicago, standard—or standards based on other carbonates with a known relationship to PDB, e.g., Nier–Solenhofen limestone—is widely employed as the zero reference point for mass spectrometric measurements of $^{13}C/^{12}C$ ratios in the same manner that NBS oxalic acid standards are directly or indirectly used as the reference standards for ^{14}C measurements. Stable carbon isotopic values ($^{13}C/^{12}C$) are expressed in per mil (per thousand) rather than percent (per hundred) deviation from a standard. For example, a $^{13}C/^{12}C$ ratio expressed as -10 per mil (‰) with respect to (wrt) PDB indicates that a sample contains 1 percent (%) less ^{13}C than does the PDB standard. Another way to express it is to say that this sample is "lighter" or depleted in ^{13}C by 10 ‰ with respect to the standard. A $^{13}C/^{12}C$ ratio is expressed in terms of a $\delta^{13}C$ value as defined in (4.7) (cf. Olsson and Osadebe, 1974):

$$\delta^{13}C \ (‰) = \left(\frac{^{13}C/^{12}C_{sample} - {}^{13}C/^{12}C_{std}}{^{13}C/^{12}C_{std}} \right) 1000 \qquad (4.7)$$

Using the same symbols for the net activity of a contemporary ^{14}C standard A_0 and unknown-age sample A, expression (4.8) defines a $d^{14}C$ value. Since $\delta^{13}C$ values are expressed in *per mil* units, $d^{14}C$ values are also expressed in per mil equivalents:

$$d^{14}C \ (‰) = [(A/A_0) - 1]1000 \qquad (4.8)$$

To normalize $d^{14}C$ onto a common $\delta^{13}C$ scale, a $D^{14}C$ value can be calculated. By general agreement (Section 4.3.2), conventional ^{14}C age estimates are normalized to a value of $-25\ \%_0\ \delta^{13}C$ wrt PDB (Ralph, 1971:21; Stuiver and Polach, 1977). A $D^{14}C$ value can be calculated using the expression in (4.9):

$$D^{14}C(\%_0) = d^{14}C - 2(\delta^{13}C + 25)(1 + \frac{d^{14}C}{1000}) \qquad (4.9)$$

A $D^{14}C$ value can be used to derive a conventional ^{14}C age employing the expression in (4.10) (cf. Gillespie, 1984:21–22):

$$t\ (\text{years}) = 8033\ \ln\frac{1}{1 + D^{14}C/1000} \qquad (4.10)$$

Because of what was perceived as the increasingly complex character of deriving accurate ^{14}C age estimates (due to the use of two half-lives and various correction and calibration procedures), one writer (Ewer, 1971) suggested the use of a measurement unit—he suggested calling it a "Libby"—that would directly reflect the original laboratory analyses used to derive a ^{14}C age value in year units. Although the specific suggestion was never adopted by any representative group of radiocarbon specialists, the $D^{14}C$ value approaches the original conception of a "Libby."

Table 4.3 illustrates the process of converting net count rates into ^{14}C age estimates over the age range from 0 to over 50,000 ^{14}C years. The net count rate of a sample exhibiting a ^{14}C age of 0 B.P. has been set in our hypothetical counter for convenience at 100 counts per minute (cpm). This value would be the count rate of a contemporary standard corrected for fractionation, Suess and/or Atomic Bomb effects. For example, if the old NBS oxalic acid standard were used, 95% of its normalized net count rate would be equal to 100 cpm under a set of standard experimental conditions. The count rates listed in column 2 would be obtained in the same detector system under the same experimental conditions corrected for fractionation in the sample. The actual values that would appear in columns 1 and 2 would be unique for each counting system. However, the ratios and other values listed in the remainder of Table 4.3 will remain constant. Note that a count rate one-half that of the contemporary standard (which yields an A_0/A ratio of 2 and is equal to a $D^{14}C$ value of -500 per mil) has an age equivalent in years of one half-life. Using the Libby value, the age assigned to such a sample would be 5570 years. Radiocarbon age values are typically rounded depending on the magnitude of the standard error associated with them. For a standard error of between ±50 and ±100, the typical Holocene ^{14}C age value is now generally rounded (as in t_2) to the nearest multiple of ten (Stuiver and Polach, 1977:362). This was not the practice initially,

TABLE 4.3
Relationship between A_o/A, $D^{14}C$, and Derived ^{14}C Age $(t)^a$

A_o (cpm)	A (cpm)	A_o/A	ln A_o/A	t_1^b (years)	t_2^c (years)	$D^{14}C$ (‰)
100	100	1	0	0	0	0
100	75	1.33	0.287	2310	2310	−250
100	50	2	0.693	5568	5570	−570
100	36.7	2.72	1.000	8033	8030	−630
100	25	4	1.386	11,136	11,140	−750
100	10	10	2.302	18,496	18,500	−900
100	5	20	2.995	24,064	24,100	−950
100	1	100	4.605	36,993	37,000	−990
100	0.1	1000	6.907	55,489	55,500	−999

^{a}5568 (5570)-year half-life (8033 mean-life) is used. Assume $\delta^{13}C$ of $-25‰$ wrt PDB.
$^b t_1$ is obtained by multiplication of ln A_o/A by the mean-life, 8033 years.
$^c t_2$ is the rounded value of t_1.

as, for example, can be seen by consulting Libby's lists of dates (e.g., Libby, 1952a:70–96; 1955:77–140).

4.7 STATISTICAL CONSTRAINTS

A ^{14}C age determination is obtained by measuring the amount of ^{14}C contained in a sample and comparing that value against the ^{14}C concentration in an appropriate standard. Since in both decay and direct counting one can never measure all of the ^{14}C contained in a sample or standard, it is necessary to consider the statistical constraints that define the precision of a measurement. All appropriately documented ^{14}C age values are cited in a format that expresses the calculated age along with an estimate of the experimental or analytical precision. The view of the journal *Radiocarbon* is that this value should express the laboratory's estimate of the precision of the measurement "as judged on physicochemical (not geologic or archaeologic) grounds." In most cases, it can be assumed that the estimate of the analytical or experimental precision will be dominated by counting statistics. Traditionally, such "±" values are calculated based only on the measurement of the activity of the sample, background, and contemporary standard. However, in some laboratories, the standard error value is an estimate of the overall sample reproducibility and takes into account other factors in addition to counting statistics (e.g., Walker and Otlet, 1985; Otlet, 1979).

In measuring the decay rate of any ^{14}C sample, an investigator is confronted with the physical fact that radioactive decay is a random process. This means that there is no way of knowing when an individual ^{14}C nucleus will decay. However, if one monitors the decay of a large number of ^{14}C nuclei over a relatively long counting period, a pattern begins to emerge. With the counting periods broken into equal time intervals, the distribution of the number of decay events in the intervals should approximate what statisticians call a *normal distribution* in which the different interval values are clustered around the average in a roughly symmetrical fashion. It is assumed that if the decay events are truly random, if no other factors intervene, and one had the time to measure an infinite number of events, the counting data would be identical to a normal distribution. Obviously, since one never has an infinite amount of time, this can never be achieved. Thus the need exists to apply statistical concepts to accurately represent the precision of the physical measurements actually obtained (Friedlander *et al.*, 1964:166–190).

Figure 4.15 illustrates a normal distribution curve superimposed over an actual set of counting data. This type of plot represents the fact that, in normally distributed data, about 68% (about 2 out of 3) of the separate measurements of the count rate should not deviate more than one standard deviation, or *one sigma*, (1σ) from the average count rate. Likewise about 95% of the time, the rates should fall within two standard deviations (2σ) of the average count rate, and about 99% of the time they should fall within three standard deviations (3σ) of the average rate. By consensus (e.g., Stuiver and Polach, 1977:557) conventional ^{14}C age estimates are cited in the form: age estimate (calculated in a standard manner) ±1σ. In most cases, the " + " and " − " values of the standard deviation are represented as being equal, e.g., ±80 years. Strictly speaking, however, because of the nonlinear character of radioactive decay, these values are not of exactly equal magnitude. However, the difference becomes significant only for samples exhibiting very low count rates, i.e., relatively

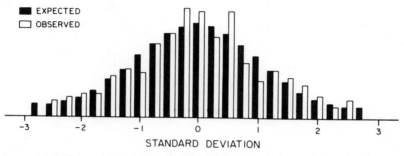

Figure 4.15 Comparison of normal distribution curve with actual counting data.

old samples. In these cases, such samples are sometimes assigned an *asymmetric* standard deviation expression, e.g., Ly-1988, 30,600 $^{+1600}_{-1300}$ (Evin *et al.*, 1983:77). For most laboratories, the value of 1σ is based on the combined statistical deviations observed in the count rates of the sample, background, and contemporary standard. However, in some facilities, variations in experimental conditions (e.g., voltage, barometric pressure, or temperature corrections) are also included when calculating statistical deviation values. Also, several different mathematical expressions are employed to calculate counting errors. As a consequence, what is represented by the ±1σ expression may vary somewhat from laboratory to laboratory. Analysis of inter- and intra-laboratory data that are published points to the fact that the 1σ expression typically underrepresents actual experimental variations. In fact, it has been suggested that *doubling* that stated one sigma value yields a more realistic estimate of actual overall interlaboratory measurement variability (International Study Group, 1982; Scott *et al.*, 1983; cf. Currie and Polach, 1980; Otlet *et al.*, 1980; Sheppard *et al.*, 1983). For this reason, it is proposed that the 1σ value of ±80 years (±1%) be considered as a *minimum* value in routine comparisons of ^{14}C age estimates (cf. Section 4.4). This minimum value may be lowered on a case-by-case basis for analysis carried out by laboratories producing documented high-precision ^{14}C values.

CHAPTER 5

EVALUATION OF RADIOCARBON DATA

5.1 GENERAL PRINCIPLES

The accuracy of an individual ^{14}C determination is directly related to the degree to which the assumptions of the ^{14}C method are fulfilled for the carbon-containing materials in a given sample. Unfortunately, it is often difficult to evaluate directly the various factors that could influence the accuracy of a single ^{14}C value. For this reason, little reliance should be placed on an individual ^{14}C "date" to provide an estimate of age for a given object, structure, feature, or stratigraphic unit. A critical judgement of the ability of ^{14}C data to infer actual age can be best made with a suite of ^{14}C determinations on multiple samples drawn from the same context or with multiple ^{14}C determinations obtained on different fractions of the same sample (Waterbolk, 1983b:18, 1971:19). Concordance of values on different sample types or fractions of the same sample from well-defined stratigraphic contexts provides one of the strongest arguments for the accuracy of age assignments based on ^{14}C values. Unfortunately, until recently, multiple ^{14}C analyses on different fractions of single samples have not been routinely employed due to limitations on sample size and costs. However, with the increasing use of AMS counting (Section 4.5), it is to

105

be expected that analysis of multiple components of samples will in the future become increasingly common (Mook, 1984).

In a critical, systematic evaluation of ^{14}C data from an archaeological perspective, it is helpful to consider those factors that can affect the accuracy and precision of a set of ^{14}C determinations. In this discussion, *accuracy* involves the correctness of the age assignment, i.e., how close the ^{14}C age estimate is to the actual or true age of the event or phenomenon in absolute terms. *Precision* refers to the time range within which the true age of the event or phenomenon is thought to lie, i.e., the geochemical/ geophysical and/or statistical characteristics of the "time envelope" being assigned. Strictly speaking, precision refers to the overall reproducibility of results, the range in values that would be obtained on measurements on duplicate samples (cf. Topping, 1962:14).

In practical terms of relevance to archaeological issues, what constitutes a relatively "high" or "low" degree of precision will vary with the quality of the understanding of chronological relationships for a particular past society or technological tradition. For example, to state that an event occurring in Egypt in the year 2850 B.C. can be dated to the early third millennium B.C. is an accurate age estimate but it might be considered by an Egyptologist to lack precision. By contrast, a statement that the same event occurred in the year 1534 B.C. would lack accuracy but was expressed with high precision (i.e., ±1 year). In cases where chronological relationships are relatively well-documented by, for example, textual data, precise age estimates are those expressed in single year and decade units. In such cases, ^{14}C data obviously cannot be used to produce precise age estimates. In nonhistoric archaeological contexts, century increments would constitute very precise age estimates. For many archaeological or paleoanthropological (and geological) situations, particularly in the Pleistocene, significant conclusions can usually be drawn if chronological relationships can be constructed by ^{14}C data with much larger temporal increments.

We have distinguished four factors that can influence the accuracy and precision of ^{14}C determinations (Section 2.1). These include

(i) *sample provenance factors* involving the integrity of the association of a sample with an event or phenomenon for which temporal placement is sought,

(ii) *sample composition factors* primarily relating to contamination and fractionation effects,

(iii) *experimental factors* including such issues as the implications of the statistical nature of ^{14}C counting data, and

(iv) *systemic factors* involving reservoir and major trend variations/de Vries effects.

A critical, careful evaluation of a set of ^{14}C determinations involves attention to how these factors separately and together might affect a given ^{14}C value and how correction and calibration procedures might be appropriately applied.

Table 5.1 provides a brief checklist of major causes of anomalous age estimates divided into those factors that would cause (I) "younger than expected" and (II) "older than expected" values. This obviously is not an exhaustive listing, but taken together these factors probably explain a significant percentage of the problematical results. Heading each list are problems with either misidentification of sample with context and/or unidentified disturbance of depositional conditions. A second set of causes involves insufficient laboratory sample pretreatment efforts. A third grouping would involve reporting errors (e.g., mislabeled samples in the field or laboratory). Such listings of potential sources of anomalous results emphasizes again the need for suites of ^{14}C values to provide checks against the single potentially problematical ^{14}C age estimate.

TABLE 5.1
Major Sources of Anomalous ^{14}C Values for Typical Archaeological Contexts[a]

I. Apparent age significantly younger than expected
1. Misidentification of sample with stratigraphic level or purported context.
2. Reworked or eroded deposits, mixing of deposits by bioturbation or geoturbation.
3. Insufficient removal of rootlets (from charcoal and bone).
4. Insufficient removal of organic decay products (humics) derived from stratigraphically higher levels.
5. Inappropriate application of reservoir correction values.
6. Careless sample storage or inappropriate sample containers (paper bags, cloth bags, cardboard boxes).
7. Unreported application of preservative produced from modern carbon source.
8. Mislabeled samples.

II. Apparent age significantly older than expected
1. Misidentification of sample with stratigraphic level or purported context.
2. Reworked or eroded deposits, mixing of deposits by bioturbation or geoturbation.
3. Insufficient removal of organic decay products (humics) derived from older deposits.
4. Mixture with fossil carbon source (i.e., asphalt, tar, lignite, coal).
5. Insufficient removal of groundwater carbonates.
6. Inappropriate application of reservoir correction values.
7. Unreported application of preservative prepared from fossil carbon source.
8. Mislabeled samples.

[a]After Terasmae (1984:10).

5.2   SAMPLE PROVENANCE FACTORS

Careful documentation of the archaeological, historical, and/or geological context of sample material is of primary importance in a critical
evaluation and utilization of ^{14}C data. The most careful analytical work
will not overcome the problem of samples collected without sufficient
regard for the problem of context and association. The most exacting attention to detail in laboratory procedures cannot ensure an accurate temporal assignment for archaeological or historical events in the absence of
an unambiguous and direct relationship between sample and event or phenomenon for which temporal placement is sought (cf. Matson, 1955:162–
163; Watts, 1960:116; Griffin, 1965:123; Stuckenrath, 1965; Pardi and
Marcus, 1977). The great variety of depositional conditions and techniques
of recovery of sample materials from archaeological sites makes it essentially impossible to provide a rigid framework that would apply in every
situation. A small set of guidelines and cautionary notes would be more
appropriate. The following comments in this section are based largely on
suggestions contained in Taylor (1970), Dean (1978), Waterbolk (1971,
1983a, 1983b, 1983c) and Nydal (1983a).

The contextual elements of a critical utilization or evaluation of ^{14}C
data involve, first, the specific delineation of the nature of the event, phenomenon, or object for which temporal placement is being sought, and,
second, the identification of the nature of the relationship or association
between an event/phenomenon and sample material(s) to be used for the
^{14}C analysis. The basic principle was set forth more than two decades ago
by Frederick Johnson (1965:776) when he suggested that a ^{14}C age estimate
" . . . does not date a site or building, or a grave or a level. The date is
that of the sample and it is the task of the archaeologist to discover the
true relationship between the sample and the area or place it came from"
(cf. Stuckenrath, 1965; Davis, 1965). Although there are limited published
data to support this assertion, it is the view of the author that the cause
of the majority of seriously anomalous ^{14}C values is a misassociation or
misidentification of sample context or provenance (cf. Polach and Golson,
1966:4; Smith *et al.*, 1971:102; Sheppard, n.d.:5). An example of the problem of stratigraphic misattribution can be exemplified in the first ^{14}C age
determinations that related to the disputed question of the timing of the
earliest arrival of human populations in the Western Hemisphere.

The initial list of ^{14}C values issued by Arnold and Libby included the
result of a measurement on a sample from the Folsom site in New Mexico.
The sample (C-377) was initially described as charcoal from a fire pit situated below bison bones and artifacts collected by H. J. Cook in 1933.

The Chicago ^{14}C age assigned to this sample (C-377) was 4283 ±250 years (an average of two determinations), which generated the comment "surprisingly young" (Arnold and Libby, 1950:10). Cook revisited the Folsom site in June 1950 and determined that the "sample had been taken from a hearth in the fill of a secondary channel which had cut through the original deposit of bison bone and artifacts" (Roberts, 1951:116). In the first formal publication of the results (Arnold and Libby, 1951:116), C-377 was listed as charcoal from a "hearth in secondary channel of later date than bison and artifact deposit." A ^{14}C value of 9883 ± 350 (C-558) was subsequently obtained on burned bison (*Bison antiquus*) bone from what was interpreted as the Folsom horizon at Lubbock Lake, Texas (Libby, 1951:293). This value according to Roberts "more closely approximates the magnitude estimated for Folsom on geologic evidence" (Roberts, 1951:20–21, cf. Haynes 1982:384). More recently, geological evidence combined with additional ^{14}C data points to the conclusion that the burned bone sample used for C-558 did not, in fact, come from the Folsom levels at the Lubbock Lake site (Holliday and Johnson, 1986; cf. Haas *et al.*, 1986). If this is correct, the first ^{14}C age determination actually associated with Folsom materials was obtained on charcoal (I-141, 10,780 ± 375 ^{14}C years B.P.) collected at the Lindenmeier site in Colorado (Haynes and Agogino, 1960).

The series of reinterpretations of the context of the ^{14}C-dated material presumably associated with the Folsom horizon provides an excellent illustration of the consequences of an incomplete documentation of sample context. It also highlights the sometimes critical role that a geologist familiar with local Quaternary sediments can play in the interpretation of ^{14}C data. Another illustration of this process involves the evaluation of temporal placement of two purported bone tools excavated from late Pleistocene sediments at the Tule Springs site in Nevada. Their initial assignment of age, based on ^{14}C determinations on presumably associated charcoal samples, was in the 20,000- to 40,000-year range. However, detailed geologic and geochemical studies, as well as a large suite of ^{14}C values, provided evidence that the bone objects were associated with sediments no older than 13,000 ^{14}C years B.P. (Haynes *et al.*, 1966).

In some contexts, the expertise of a Quaternary geologist would also be helpful to evaluate the degree of temporal homogeneity of sample materials in relationship to a depositional feature for which time placement is being sought. This is true, for example, in situations where very small charcoal fragments recovered from fluvial (water-laid) sediments are employed. In one study, variations in ^{14}C age of as much as 1000 years were correlated with variations in particle size of charcoal fragments taken from a modern surface bulk soil matrix. In this case, increasing age was associated with decreasing particle size in the range of 8 to 0.5 m (Blong and Gillespie, 1978). Such a range in ^{14}C values reflects the "residence time" for particulate charcoal in these sediments. When charcoal fragments in the millimeter size range from fluvial environments are to be used for ^{14}C age estimates, it would seem to be prudent to first evaluate the degree of temporal homogeneity exhibited by such samples. Several instances of very large (>20,000 years) deviations between small charcoal fragments and other sample types from the same stratigraphic levels in alluvial depositional contexts point to the need to exercise great caution in the interpretation of the ^{14}C data from such samples (Evin *et al.*, 1983:77).

In regions where soils are affected by various frozen-ground phenomena (e.g., permafrost), severe depositional discontinuities caused by various *geoturbation* processes have been identified that make it dangerous to employ ^{14}C determinations obtained on "associated" organics to infer age for archaeological materials. In circumpolar regions, vertical and horizontal disturbances in the soil profiles can be so serious that one observer suggested that a ^{14}C determination "obtained from a sample found in direct association with an artifact may have absolutely no relationship to the age of the artifact regardless of the stratigraphic appearance of the containing earth" (Campbell, 1965:184). It might also be noted that in environments where wood is both rare and preserved on ground surfaces over long periods (e.g., arctic and subarctic regions), the use of such materials for fuel in camp fires has been shown to introduce significant errors in assigning age to particular stratigraphic levels or associated tool types. Also, in areas where surface or near-surface fossil fuel materials (e.g., tar or bitumen and various types of coals) were exploited by aboriginal groups, the possible incorporation of these materials into a sample matrix must be carefully evaluated (Evin, 1983).

The loss of contextual integrity in archaeological deposits can be critical in cases where, for example, age estimates for specific paleobotanical remains, particularly involving early occurrences of domesticated or cultivated plants, are made on the basis of ^{14}C values obtained on presumably "associated" organics such as charcoal. The documentation of the relationship between the materials on which the ^{14}C analysis has been obtained and the paleobotanical evidence must be particularly conclusive when the estimates are used to make inferences that represent a significant departure from generally accepted views. An example of such a problem arose as a result of the suggestion that food production involving the use of domesticated barley (*Hordeum vulgare*) and einkorn wheat were being utilized by Late Paleolithic populations near Aswan in Egypt about 17,000–18,000 ^{14}C years ago, nearly 10,000 years earlier than previously documented in the Near East. The assignment of age to these paleobotanical specimens was based on what appeared to be a good association in a buried hearth. Radiocarbon analysis carried out by AMS methods on individual barley seeds from the site determined that none exhibited ages in excess of 5000 ^{14}C years, dating them well after the beginning of the use of cultivated plants in other areas of the Near East (Wendorf *et al.*, 1979, 1984).

In another study, AMS ^{14}C analysis of carbonized fragments of squash (*Cucurbita* sp.) confirmed its occurrence in Archaic period deposits in Illinois at about 7000 ^{14}C years B.P. By contrast, AMS analysis of fragmentary remains of maize (*Zea mays*) previously dated at about 2000 ^{14}C years on the basis of associated organics determined that the actual age of the maize itself was about 1500 ^{14}C years B.P. at one site and less than 600 years at three other localities in the region. In one case, the maize fragments were determined to be modern contamination (Conrad *et al.* 1984). AMS based ^{14}C determinations were also used to document conclusively the indigenous occurrence of a North American plant thought by some to have been introduced at the time of European contact. Individual seeds of *Corispermum* L. were analyzed to eliminate the problems of stratigraphic mixing and late Pleistocene/early Holocene ages were obtained on four specimens (Betancourt *et al.*, 1984).

In some cases, the loss of stratigraphic association for samples can be traced to disturbance caused by human and animal activity, which can occur both during occupation as well as after the abandonment of sites. Such *anthropogenic* and *zoagenic bioturbation* processes can cause dating anomalies in many types of sites. It can be particularly severe in cave and rock shelters that contain organic materials of significantly different ages. An example of such a problem is illustrated by the ^{14}C analysis of materials from Gypsum Cave, Nevada. In 1931, atlatl shaft fragments and dung from an extinct giant ground sloth were found in apparent association. The age of the dung was determined to be approximately 10,000 years old (10,445 ± 340 ^{14}C years B.P., C-221). This value was used to infer an age for the atlatl fragments (Libby, 1952a:85). More than a decade later (Berger and Libby, 1967:480), a ^{14}C analysis of one of the atlatl shafts determined its age to be about 3000 years B.P. (2900 ± 80 ^{14}C years B.P., UCLA-1223). The same difficulty was encountered in determining the age of fragments of atlatl dart shafts that had been recovered from Potter Creek Cave in northern California in the early part of this century. The dart shaft fragments were originally thought to have been contemporaneous with extinct Pleistocene fauna found in the cave. The age of the atlatl shaft fragments was later determined to be about 2000 years (1915 ± 150 and 1910 ± 150 ^{14}C years B.P., UCR-148 and UCR-151), indicating at least a

6000–8000-year minimum temporal hiatus between the extinct fauna and the cultural materials (Taylor, 1975; Payen and Taylor, 1977).

The degree of vertical mixing can sometimes be inferred on the basis of geomorphological or archaeological evidence. However, conclusions based on evidence from these sources are often equivocal. A more definitive, quantitative approach would be to analyze a series of carefully selected samples to test for depositional mixing. An example of where this was carried out is illustrated in Fig. 5.1. Radiocarbon determinations were carried out on wood, charcoal, and carbonized grain samples from a historically well-dated (*ca.* A.D. 600–A.D. 1200) stratified Medieval fortified village in northern Germany. Excavation had identified five major periods of fortification construction. Comparing the ^{14}C data with the known-ages of the levels from which the samples were obtained revealed a very poor correlation. The range in ^{14}C values exhibited by multiple samples from each depositional unit was about 300–500 years, i.e., almost the entire period of the site occupation. Examination of the excavation

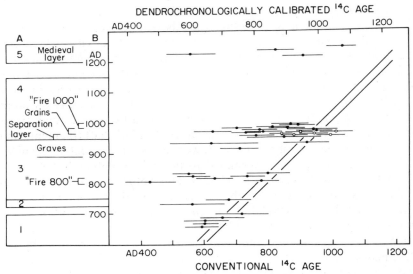

Figure 5.1 Comparisons of archaeological and ^{14}C ages from Oldenburg, Schlewig-Holstein, Germany (Willkomm, 1983:646). Solid circles, wood and charcoal; open circles, carbonized grains. A, fortification levels; B, archaeologically inferred ages. Diagonal lines indicate exact agreement between age based on archaeological evidence and ^{14}C values. Upper diagonal is for dendrochronologically calibrated ^{14}C values. Lower diagonal is for conventional ^{14}C values.

records confirmed that every stratigraphic level contained materials from the beginning of the settlement due to frequent burning and reconstruction of the fortifications (Willkomm, 1983).

Another example where a suite of ^{14}C determinations confirmed observations that the stratigraphic integrity in a site had been compromised can be illustrated from data obtained at Gombe Point, near Kinshasa in Zaire (Cahen et al., 1983). Since the late 1920s when the site was initially excavated, the Gombe Point sequence had provided one of the principal sources for the dating of the late Pleistocene and Holocene lithic industries of Central Africa. Excavations conducted in the early 1970s observed that lithic specimens that could be refitted back to a single core were, in many cases, recovered at significantly different depths in the deposit. On this basis, these lithic samples would have been dated to different periods (Cahen and Moeyersons, 1977). A series of ^{14}C determinations on charcoal samples derived from features with presumed associations with the various lithic "industries" showed no consistent age attributions. Samples from a single 2.5 m trench yielded several inversions in ^{14}C values (cf. Ashmore and Hill, 1983).

When major anomalies are present in ^{14}C data, a listing of possible explanations based on conditions at the site would be helpful even if it might be difficult or impossible to demonstrate conclusively the suggested cause(s) at the site originally involved. Future investigators would then be alerted to the potential problem and might be able to collect samples and other data that would permit an evaluation of the proposed explanation(s). For example, excavators collecting samples in caves or rock shelters containing packrat nests might be made aware of the possibility that aboriginal populations may have employed packrat midden material as sources of fuel in hearths. It has been determined that in some cases, wood and other organics spanning as much as 30,000 to 40,000 years can be contained in such midden accumulations. Samples of very different ages could be distributed in a deposit if a packrat nest was disturbed in any way. This possibility was raised by excavators at Pintwater Cave, Nevada, when charcoal from a hearth feature associated with cultural materials which should date at about 2000 years B.P. (Basketmaker) yielded an age of about 9000 ^{14}C years B.P. Pintwater Cave contained a number of packrat middens. The suggestion was that a "lazy Basketmaker" robbed one or more packrat nests for fuel (Haynes, 1965:149–150).

Dean (1978) has distinguished three types of "event time" that might be considered in evaluating ^{14}C data in some contexts: target event E_t, reference event E_r, and dated event E_d. E_t is the behavioral or archaeological expression for which temporal placement is being sought; E_r is the closest datable event that can be related to the E_t, and E_d is the event actually dated by ^{14}C analysis. Several types of discontinuities between E_d and E_t can be defined. These terms are defined by illustration in Fig. 5.2. A *disjunction* exists when E_t lies later in time than E_d; a *disparity* occurs when the reverse is true. A discontinuity between E_t and E_r is labeled a *hiatus* and between E_r and E_d a *gap*.

In many instances, it is not possible to quantitatively determine the magnitude of a disjunction or disparity. In these cases, at least some estimate of the maximum and minimum ranges might be ascertained. Table 5.2 provides a complementary framework within which consideration can be given to the relationships of E_t and E_d. In cases were the ^{14}C analysis is performed on the object for which temporal placement is sought (i.e., $E_d = E_t$), there is essential certainty in the association of sample with archaeological expression. As the relationships become less direct, the level of confidence will, of course, decrease. As far as problems of provenance are concerned, the number of separate ^{14}C estimates that are appropriate will be inversely proportional to the confidence level; as the confidence level is reduced, the number of independent ^{14}C analyses needed increases. One of the important consequences of the increased use of AMS direct counting methods (see Section 4.5) on a routine basis will be the ability to provide more "A1" (cf. Oakley, 1966) type ^{14}C age estimates for archaeological samples.

Another aspect of contextual concerns relates to the use of ^{14}C data outside the task of deriving chronology. Dean (1985), for example, has noted the inappropriate use of the distribution of ^{14}C values to estimate the relative intensity of occupation in a region. This would, in most cases, be very problematical since many circumstances can influence these

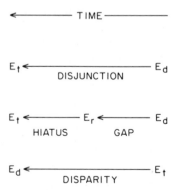

Figure 5.2 Definitions of relationships between various event times based on Dean (1978). E_t, target event; E_d, dated event; E_r, reference event.

TABLE 5.2
Confidence Levels Defining Reliability of Association of Sample Material with Archaeological Feature

	Confidence level	Defintion	Example
1[a]	Essential certainty[b]	[14]C analysis on object for which temporal placement sought	[14]C analysis on human bone to obtain age estimate on burial
	High probability	[14]C analysis on organics in direct *functional* relationship with object/event for which temporal placement sought	[14]C analysis of textile used to wrap burial to obtain age estimate on burial
2	Reasonable possibility	[14]C analysis on organics in enclosing deposits of assumed similar age to the object/event for which temporal placement sought	[14]C analysis of charcoal in sediments adjacent to burial to obtain age estimate on burial
3	Possibility	[14]C analysis of organics from deposits correlated with deposits containing object/event for which temporal placement sought	[14]C analysis of charcoal in sediments associated stratigraphically/geomorphologically or cross-dated on the basis of some cultural feature with the burial to obtain age estimate on burial

[a]Based on Oakley (1966:2–8). Cf. Mann and Messman (1976).
[b]Modified from Waterbolk (1971:15–16; 1983a:58).

numbers, physical preservation of sample materials probably being the major factor.

5.3 SAMPLE COMPOSITION FACTORS

Chapter 3 reviewed the range of sample types typically recovered in archaeological contexts on which ^{14}C measurements can be obtained and briefly noted the nature of the various contamination and fractionation factors that should be considered in evaluating ^{14}C values. This section will provide a general perspective on the *magnitude* of expected contamination and fractionation effects. This is particularly important in the case of potential contamination of samples, since the discussion of apparently anomalous ^{14}C values has often employed this as an explanation.

5.3.1 Contamination Effects

In order to put the issue of contamination in an appropriate context, it might be helpful to first review the result of the introduction of various amounts of known-age contaminants into a sample of known age. The most dramatic effects result from the addition of contemporary or modern ^{14}C to samples of Pleistocene age (>10,000 years B.P.) particularly for those materials that have finite ages greater than about 40,000 years. For example, the addition of 1% modern ^{14}C to a sample exhibiting a finite ^{14}C age of about 50,000 years will result in an apparent age of about 35,000 years. Figure 5.3 illustrates the reason for this approximately 15,000-year reduction in age. At this part of the ^{14}C decay curve, only a few tenths of a percent change in ^{14}C activity translates into an age difference of thousands of years. The relationship between the actual and apparent age of samples to which varying percentages of modern carbon have been added is presented in Fig. 5.4. Note that relatively high levels of modern contamination provide a limit to the maximum resolvable ages possible. For example, samples with ages in excess of about 45,000 years with 5% modern ^{14}C contamination will exhibit a maximum composite apparent age of only about 25,000 years regardless of the true age of the material.

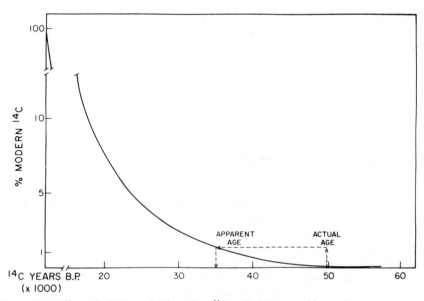

Figure 5.3   Effect of addition of 1% modern ^{14}C to 50,000 year old sample.

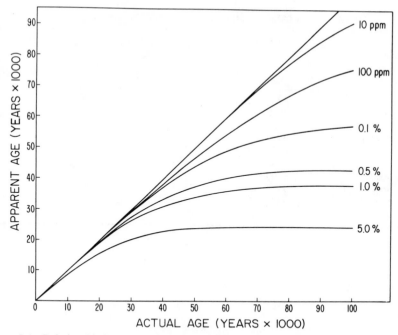

Figure 5.4 Relationship between actual age and apparent age for samples to which varying percentages of modern carbon have been added. [From Taylor (1982).]

In Fig. 5.4, the effect of modern ^{14}C contamination has been extended out to 100,000 years in anticipation of the utilization of direct counting by AMS techniques for samples with ages in excess of 50,000 years (Section 4.5). For samples in this extended age range, it is clear that modern carbon contamination must be kept below about 10 parts per million (ppm). Only with the most vigorous pretreatment and sample handling techniques can the practical potential of AMS technology in this age range be achieved. By contrast, the effect of the contribution of carbon of infinite ^{14}C age (e.g., coal, lignite, or carbonates from sedimentary limestone deposits) to recent samples is much less severe. For example, the introduction of 1% "dead carbon" to a modern sample will result in an increase in age by about 80 years. A 10% contribution will result in an apparent age for this sample of about 850 years.

The vast majority of samples submitted for ^{14}C analysis by archaeologists are of Holocene age. The effect of contamination with modern ^{14}C from 1% to 20% is presented in Table 5.3. For example, a 1% addition of modern ^{14}C to a sample with an actual age of 5000 years will result in

TABLE 5.3
Effect of Introduction of Modern Carbon to Samples of Holocene Age

Actual age (years B.P.)	Approximate age with following percentage addition of modern carbon			
	1%	5%	10%	20%
1000	990	950	900	800
2000	1950	1890	1750	1550
3000	2950	2800	2650	2300
4000	3950	3750	3500	3000
5000	4950	4650	4350	3700
6000	5900	5550	5150	4400
7000	6900	6450	5950	5050
8000	7850	7350	6750	5650
9000	8850	8200	7500	6200
10,000	9800	9050	8200	6800

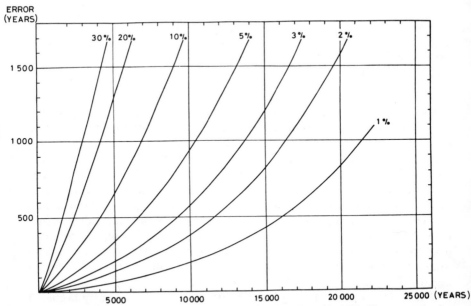

Figure 5.5 Effect (in years along ordinate) of the introduction of 1–30% older contamination (lower ^{14}C activity) than sample to be dated. Values along the abscissa indicate difference between the true age of sample to be dated and older contaminant. [From Olsson (1974).]

an apparent age of about 4950 years; a 5% addition would result in about a 350-year decrease. The effect of similar amounts of modern contamination increases with the actual age of the sample. The 350-year anomaly for the 5000-year-old sample increases to almost 1000 years for a sample 10,000 years old contaminated with the same 5% modern carbon. More typical, however, will be situations in which carbon-containing compounds differing in age from the original sample material from a few hundred to a few thousand years are physically or chemically incorporated into a sample. Ingrid Olsson of the University of Uppsala (Sweden) Radiocarbon Laboratory has prepared two very helpful plots that illustrate the amount of error introduced into sample ages as the result of the introduction of different percentages of contaminants of younger (Fig. 5.5) or older (Fig. 5.6) age than the sample material (Olsson, 1974a; cf. Broecker and Kulp, 1956).

Such tables and plots point to the fact that if the difference in age between a sample and any contaminant does not exceed about 5000 years and is limited to 1–2%, the errors introduced from contamination will

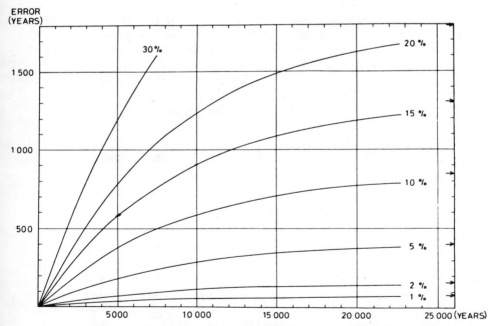

Figure 5.6 Effect (in years along ordinate) of the introduction of 1–30% younger contamination (higher ^{14}C activity) than sample to be dated. Values along the abscissa indicate difference between the true age of sample and the younger contaminant. [From Olsson (1974).]

generally not exceed a few hundred years. Appeal to contamination as an explanation of seriously anomalous ^{14}C values must conform to the strictures as exemplified in the data presented in Table 5.1 and Fig. 5.4 through 5.6. For example, one laboratory reported that a sample estimated as being 5000 years old yielded a ^{14}C value of about 3600 years B.P. It is interpreted as being 1400 years too young because of recent carbon contamination (Delibrias *et al.*, 1974:20). In such a case, more than 20% of the sample would have to be modern material. This degree of contamination should be rare. Sample pretreatment by the majority of ^{14}C laboratories is generally very efficient in removing most of the contaminating organics from sample types usually recovered from archaeological contexts. It is often not possible to remove *all* of the non-*in situ* organics from some samples, but in the vast majority of cases any remaining contamination in Holocene age samples will not be sufficient to alter the indicated age of the sample by an amount in excess of the statistical variance attached to the age estimate.

It seems reasonable to suggest that specific geochemical evidence be offered when "contamination" is used as an explanation for an anomalous ^{14}C value. This can be accomplished, for example, by extracting various fractions of a sample and dating each separately. By this means, the nature of the contaminant and the magnitude of the dilution with younger or older material can sometimes be documented. Of the types of samples typically encountered in archaeological contexts, bone (Section 3.3.5) exhibits the greatest variability in age that can be attributed to the presence of organics of varying ^{14}C activity. For critical bone samples, particularly when dealing with samples of expected Pleistocene age, it is almost manditory that an amino acid fraction be measured. If preservation of the bone precludes such an analysis, as many other fractions need to be analyzed as possible so that a quantitative measure of contamination present in the bone can be made (Taylor, 1982).

5.3.2 Fractionation Effects

The pioneering studies of Harmon Craig (1953) pointed to the need to consider natural variations in the stable isotope ratios (^{13}C/^{12}C) of samples to obtain precise and comparable ^{14}C/^{12}C ratios. Craig's work in the laboratory of Harold Urey at the University of Chicago on natural ^{13}C variations paralleled the studies concurrently conducted by Libby on ^{14}C at the same institution. Craig showed that there had been no change in ^{13}C/^{12}C ratios, i.e., δ^{13}C values, over time. Rather, he demonstrated that there were variations in stable isotope ratios in contemporary organic materials as a function of the part of the carbon reservoir from which the samples were derived (cf. Broecker and Olson, 1959).

What effect did $\delta^{13}C$ variations have on ^{14}C values? Based on theoretical considerations and on analogies with experimental data collected for other isotopic pairs, the amount of enrichment or depletion in ^{14}C for any given carbon compound due to fractionation effects was determined to be approximately two times that measured by $\delta^{13}C$ values in the same sample (cf. Craig, 1954).[1] This suggested that samples with identical death dates would vary in their ^{14}C ages if their $\delta^{13}C$ values varied significantly. For example, marine carbonates and terrestrial wood typically differ in their $\delta^{13}C$ values by about 25‰ or by about 2.5%. This means that a marine shell and charcoal sample that should manifest the same ^{14}C age by virtue of identical death dates would, in fact, exhibit a different ^{14}C "age." In the case of the 25‰ difference in $\delta^{13}C$ values, the marine shell carbonates should exhibit a ^{14}C age approximately 400 years younger than the typical wood (two times 2.5% equals a 5% difference in ^{14}C concentration; each percent difference in ^{14}C activity equals approximately 80 years; 5 times 80 years equals an apparent anomaly of about 400 years due to the fractionation effect). Unfortunately, actual measurements of paired marine shell/charcoal or wood samples have suggested a more complex situation. Some paired samples manifest no statistically significant differences in ^{14}C age. In other cases, marine shell exhibited apparent ages of as much as 800 to 900 years. It has become clear that factors other than fractionation (and surface contamination) were responsible for such anomalies. For marine shell, it has become important to consider reservoir effects caused by upwelling of deep ocean water as discussed in Section 5.5.1.

As a result of the documentation of both terrestrial (e.g., Section 3.3.2) and marine sample natural $\delta^{13}C$ variations, a consensus has been developed that all ^{14}C values should be normalized with respect to a common $\delta^{13}C$ scale. This is particularly important when one is making comparisons of ^{14}C age determinations on a variety of organic materials that manifest a wide range in ^{13}C values. By convention, fractionation effects are evaluated by normalizing all samples to $-25‰$ with respect to the PDB standard. A conventional radiocarbon age is normalized in terms of its $\delta^{13}C$ value either directly measured or estimated (Stuiver and Polach, 1977). Any other correction such as that to deal with contamination effects or calibration procedures is then applied to the conventional ^{14}C age values following normalization onto a common ^{13}C scale.

[1]It has been argued that this factor is not well known (e.g., Radnell, 1980). However, Wigley and Muller (1981) have noted that the effect on ^{14}C values, even assuming a major variation in this parameter, would rarely exceed 20 years.

While the total range in $\delta^{13}C$ values exhibited by sample materials employed in ^{14}C dating for archaeological purposes is about -40 to $+5‰$ (with respect to PDB), the vast majority of values typically cluster between about -25 and $0‰$ (cf. Polach, 1975; Stuckenrath, 1977; Stuiver and Polach, 1977). Figure 5.7 illustrates the relationship between variation in ^{13}C values and apparent ^{14}C ages. Since a $1‰$ variation in $\delta^{13}C$ yields about a 16-year variation in a ^{14}C age estimate, the age corrections are in the 0–400 year range. Ideally, each sample should have its $\delta^{13}C$ ratio measured and the ^{14}C deduced age normalized in terms of the *measured* $\delta^{13}C$ value. When a direct measurement is not available, a $\delta^{13}C$ value may be estimated on the basis of the sample type (cf. Stuiver and Polach, 1977:358). Although some variation in the $\delta^{13}C$ values of different parts of the same plant (e.g., between leaves and wood) have been noted, the effect does not exceed $4‰$ (Leavitt and Long, 1982).

However, some sample types exhibit significant variation in ^{13}C ratios. Table 5.4 illustrates this variation for typical terrestrial organics on a worldwide basis using ^{13}C values published in *Radiocarbon* between 1970 and 1981 as compiled by Burleigh *et al.* (1984). In each case, the observed

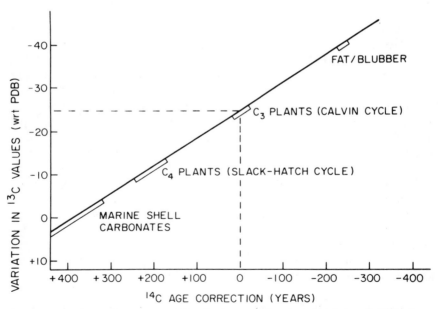

Figure 5.7 Variations in apparent radiocarbon age as a function of variations in $\delta^{13}C$ values. Amount of variation in apparent radiocarbon age determined in relationship to "normal" terrestrial organics (wood) average $\delta^{13}C$ value of $-25‰$ with respect to PDB. [Adapted from Stuckenrath (1977) and Terasmae (1984).]

TABLE 5.4
$\delta^{13}C$ Values Reported on Various Sample Materials[a]

| Sample material | $\delta^{13}C$ ‰ (wrt PDB) | | Equivalent maximum ^{14}C age range (years) |
	Mean[b]	Range	
Charcoal	-24.7 ± 1.8	-10.5 to -30.8	325
Wood	-25.2 ± 2.3	-10.1 to -31.4	340
Other plant materials	-23.2 ± 5.1	-8.7^c to -33.2	392
Bone organics (collagen), human	$-19. \pm 2.5$	-8.0^d to -24.6	266
Bone organics (collagen), nonhuman	-21.2 ± 2.7	-11.8^d to -32.8	336

[a]Based on Burleigh et al. (1984).
[b]$\pm 1\sigma$.
[c]Particularly low $\delta^{13}C$ values are reported from the Western Hemisphere and Africa.
[d]Particularly low $\delta^{13}C$ values are reported from southern Africa.

range in correction values is in excess of 250 years. Much of this variation, however, is due to the effect of mixing of C_3 and C_4 plant sources in each category. In more restricted geographical locations, the range in $\delta^{13}C$ values for a given sample type is generally much less than a worldwide average. In one study, the range in values for wood/charcoal from Norwegian sites was 6‰ for a maximum variation due to a fractionation effect of about 100 years. Bone collagen for marine mammals exhibited comparable values, but bone collagen from terrestrial animals showed a much greater dispersion of $\delta^{13}C$ values (Gulliksen, 1980). In most cases, it would be necessary to obtain $\delta^{13}C$ values on each sample if maximum precision is desired. It should be noted that when a direct measurement of the $\delta^{13}C$ value is obtained, the experimental error on this value can essentially be ignored since it typically does not exceed 0.2‰ or the equivalent of less than 4 years. If $\delta^{13}C$ ratios are not available, the precision of the ^{14}C age estimate would need to be evaluated in light of the known variability in carbon isotope fractionation effects on different types of sample materials in different localities.

5.4 STATISTICAL AND EXPERIMENTAL FACTORS

A sample for which a ^{14}C age of 5570 ± 80 ^{14}C years B.P. has been determined is *not* 5570 ^{14}C years old. (Actually, the probability that the actual age of the sample is exactly equal to the value cited as the ''age'' approaches zero.) The nature of the ''standard error'' was initially a source of misunderstanding among some archaeologists since it was assumed that the actual ^{14}C age lay within the band of time represented by the error bar (cf. Stuckenrath, 1965; Barker, 1970). The nature of the statistical variance is now generally understood, but currently there may be a degree of overconfidence that the cited counting variance represents an estimate of *overall* precision. This was probably more nearly true at a time when the statistical errors were significantly larger. With the reductions in statistical uncertainty made possible by advances in low-level counting technology, other error terms, which unfortunately are less amenable to mathematical treatment, now dominate considerations of the appropriate overall precision that can be realistically assigned to individual ^{14}C age estimates.

The expression 5570 ± 80 ^{14}C years B.P. is a convenient way of stating, using conventional parameters (Section 1.1), that the average ^{14}C activity measured in this sample was about 50% of the contemporary standard measured to a statistical precision of about ±1% (±80 years). As noted in Section 4.7, if the ^{14}C activity of this sample was measured over a sufficient length of time and the number of counts recorded were broken into equal time segments, the age equivalents for these counting rates would fall between 5490 and 5650 ^{14}C years B.P. approximately 68% of the time. Approximately 95% of the time, the equivalent age corresponding to these counting rates would fall between 5410 and 5730 ^{14}C years B.P. and about 99% of the time between 5330 and 5810. In retrospect, it might have been more helpful to have cited ^{14}C values as time intervals, as in Table 5.5, rather than employing the conventional expression. This might have avoided some of the initial misunderstandings and facilitated more appropriate comparisons between ^{14}C values.

In Table 5.5, for example, a superficial inspection would suggest that Sample 1 is 190 years older than Sample 2. This conclusion would be rapidly discarded when the values are expressed as ranges since it can be seen that there is a considerable overlap in their values at the 95% level of confidence. As suggested by Spaulding (1958), an alternative ap-

TABLE 5.5
Conventional and Interval Expression of ^{14}C Age Determinations

Sample number	Conventional expression	Probability expression		
		1 σ 68% (2 in 3)	2 σ 95% (19 in 20)	3 σ 99% (997 in 1000)
1	5570±80	5490–5650	5410–5730	5330–5810
2	5380±100	5280–5480	5180–5580	5080–5680

proach would be to apply a simple statistical principle in comparing values that contain standard errors, namely, that the standard error of the differences between two independent quantities is the square root of the sum of their squared standard errors. In our case, the difference would be 190 ± 130 (80^2 ±100^2 = 16,400; the square root of 16,400 is about 130). To evaluate the significance of this expression, one could convert it to units of standard error, in this case giving a value of approximately 1.5 (190 divided by 130 is about 1.5). In statistics texts, this value is often called t and from a table of t (for infinite degrees of freedom) the probability of obtaining such a value can be determined.

Thus, by definition, a ^{14}C "date" does not indicate a specific point in time. It expresses the time interval within which there is a given probability that the ^{14}C age equivalent of the actual ^{14}C activity of a sample actually lies. In the case illustrated above, the value of 5570 years is an artifact of the average count rate for the sample. If additional counts were recorded, the equivalent age that would be cited would be adjusted as a function of the new average count rate observed. Thus, there is a high probability that the single ^{14}C values of 5490 ± 80, 5570 ± 80, and 5650 ± 80 actually represent the same ^{14}C age since in each case the combined standard deviation exceeds the differences between each pair of values.

A good example of the effect of normally distributed counting data on ^{14}C age expressions is afforded by an experiment carried out by the British Museum ^{14}C laboratory (Barker, 1970:42). Figure 5.8 presents the results of a series of weekly measurements carried out on the same sample preparation over a period of 6 months. While the mean age value assigned to this sample would be about 4425 years (dotted line), individual measurements at weekly intervals yielded ages of 4300 (±100) to 4600 (±100) years. *It is important to note that the deviation of any weekly value from the mean value is no more than one would expect from normally distributed data.* Figure 4.15 illustrates the same principle. It should be emphasized that such experiments address *intralaboratory* variations based on counting statistics alone. Variations *between* laboratories measuring what are presumed to be duplicate samples have demonstrated that there can be as much as 4–5 times the stated 1 σ statistical variance (Polach, 1973:715). In such cases, inhomogeneous samples (a sample matrix with components exhibiting different ^{14}C activities) or variability in sample pretreatment approaches could account for some of the observed interlaboratory differences.

In addition to inherent limitations in the precision of a ^{14}C value due to the random nature of radioactive decay, there are other error factors that are less subject to analysis since they are due to undetected experimental or instrumental malfunctions or undetected human error in the laboratory. Generally, the only way in which such a situation can be detected is by analysis of suites of values in which seriously anomalous

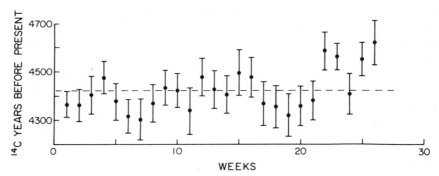

Figure 5.8 Radiocarbon values obtained from replicate measurements of a single sample preparation on a weekly basis over a 6-month period. [From Figure 1 in Barker (1970).]

determinations are identified or identical samples measured in different laboratories. A reevaluation of the experimental data can sometimes identify the problems, which can range from errors in calculations to mislabeling of samples. From time to time opportunities are afforded to publish corrected values. For example, *Radiocarbon* has published a comprehensive index of all samples published from 1950 to 1965, which provides an opportunity for laboratories to provide corrected values in cases where errors were detected (Deevey *et al.*, 1967). Less than 10% of the values published to that date required any type of correction.

5.5 SYSTEMIC FACTORS

Systemic effects on ^{14}C values are those deriving from violations of the primary assumptions of the ^{14}C method. These include the assumption of constant ^{14}C concentration in living materials over time and the fact of constant ^{14}C concentrations within each of the carbon reservoirs. Violations of such assumptions can occur within a specific geographic zone or on a worldwide basis. Correction or calibration procedures that attempt to mitigate the effect of the systemic anomalies usually assume that any effect due to problems related to sample composition (i.e., contamination or fractionation effects) has already been adjusted.

5.5.1 Reservoir Effects

One of the important features of the ^{14}C method is the potential of the worldwide comparability of ^{14}C values. For this potential to be realized, however, the *initial* ^{14}C activity of samples must have been identical, i.e., each sample must begin with the same "zero age" ^{14}C activity. If the ^{14}C activity of a *living* organism is, for example, 10% below the activity of other contemporaneous living materials, it would exhibit an apparent age of about 850 years due to a reservoir effect. The sample's indicated ^{14}C age would always reflect this 850-year anomaly.

Historically, the first materials on which reservoir effects were identified were marine shells (Section 3.3.3). After sample contamination issues were resolved and it was determined that well-preserved Holocene shells could yield carbon isotope values not affected by exchange, the contribution of other geophysical factors could be evaluated. We have already noted the fractionation effect. One approach to investigating the reservoir effect in marine shells was to examine the ^{14}C activity of contemporary samples to determine if the initial ^{14}C concentration in such materials could be significantly different from that of standard terrestrial organics. As a result, a number of measurements have been obtained on modern prebomb marine shells. *Prebomb* samples are those collected alive (or, at most, within a few years after death) before 1950–1960, i.e., before the beginning of large-scale testing of thermonuclear devices, which injected large amounts of artificial ^{14}C into the atmosphere and oceans (Section 2.5.3).

Several investigators (e.g., Mangerud and Gulliksen, 1975; Robinson and Thompson, 1981) have outlined procedures for deriving reservoir age estimates for marine shell and marine mammals. They have summarized these values for a number of regions of the world. Areas where reservoir effects have been studied in particular detail include the Pacific coast of North America (e.g., Berger *et al.*, 1966; Taylor and Berger, 1967), northern coastal Europe including England (e.g., Tauber, 1979; Olsson, 1980b; Harkness, 1983), and Australia/New Zealand (e.g., Gillespie and Polach, 1979). Figure 5.9 illustrates the type of data used to infer reservoir effects for west-facing coasts of North and South America. The procedure involves first deriving a conventional ^{14}C estimate on a modern prebomb marine sample from a known locality and collection date. A conventional age estimate takes the apparent ^{14}C age and normalizes it to a $\delta^{13}C$ value of $-25‰$ with respect to PDB (Section 5.3.2). The conventional ^{14}C age is then adjusted for the age of the sample in 1950 and, in some cases, the amount of fossil ^{14}C (Suess effect) in the oceans at the time of collection to yield a reservoir age correction for the region from which the shell was derived.

It should be emphasized that many ^{14}C laboratories do not correct for either fractionation or reservoir effects in marine shell, due to the view that, for many regions, they approximately cancel each other (cf. Stuiver, 1980:965; Olsson, 1983b). Typical fractionation effects for marine shell carbonates adjust ^{14}C values in marine shell samples by about 400 years.

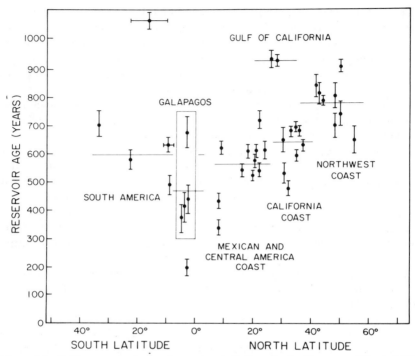

Figure 5.9 Guide to reservoir ages for marine shells from the Pacific coast of North, Middle, and South America, the Gulf of California and the Galapagos Islands. [Based on Robinson and Thompson (1981) and modified on the basis of data taken from Berger *et al.* (1966); Taylor and Berger (1967); and R. Berger, personal communication.]

This value is approximately equal to what was considered the "average" apparent age of surface ocean water based on measurements in the Atlantic Ocean (Broecker *et al.*, 1960). With increasing data, however, it has become clear that there is a significant range in reservoir values for different marine environments. Unfortunately, laboratories have not always noted their practice in deriving ^{14}C values on marine shell. Because of this, it is important to determine the custom of the laboratory that analyzes such samples.

The principal reason for the existence of these apparent ages for marine shell carbonates is the phenomenon known as "upwelling" in which water from the deeper parts of the ocean is periodically brought to the surface ("upwelled") and mixed with surface ocean water. Due to the slow mixing rates between the surface and some of the deeper portions of the ocean, the apparent ^{14}C age of some upwelled water can be in excess of 1000 years. The ^{14}C concentration of the carbonates in the surface waters is thus reduced and the marine shell reflects this activity. For this reason, the reservoir from which the marine shell derives its ^{14}C activity may not be in equilibrium with atmospheric ^{14}C activity. Upwelling is caused primarily by the effect of wind action together with the rotational force of the earth moving the surface water. The most seriously affected areas are generally on west-facing coasts and in the polar regions. However, for any particular area, the shape of the coastline, local climate, and wind patterns as well as near-shore currents and bottom topography can contribute variations in the magnitude of upwelling. In addition to the western coasts of the United States and Peru, other major coastal upwelling zones include northwest and southwest Africa and the Somalia region.

In some regions, variations in upwelling patterns can induce variations up to the equivalent of 200–300 years in the reservoir effects within relatively circumscribed areas. This can be seen most dramatically in Fig. 5.9 for the region around the Galapagos Islands off South America where contemporary shells can exhibit as much as a 400-year range in reservoir ages. Also, marine shell species whose habitat includes estuaries, bayous, inland waterways, and bay environments sometimes exhibit significant variation in their contemporary ^{14}C activity (Broecker and Olson, 1961). Living shells growing in environments influenced by the discharge of large amounts of carbonate-rich fresh water can have apparent ages of up to 1000 years. This is the case, for example, in the northern, relatively shallow, portion of the Gulf of California (Sea of Cortez) as a result of the pre-Boulder Dam discharge of the Colorado River. Since this river flows over a limestone base, ^{14}C depleted water has diluted the ^{14}C activity in the water in the upper portion of the Gulf. Marine shell growing in this area can exhibit reservoir ages of up to about 900 years (Berger et al., 1966). Figure 5.10 provides a guide for reservoir corrections for marine shell carbonates in various coastal regions of the world where reservoir effects have been studied.

Table 5.6 illustrates the difficulties in applying reservoir corrections for marine shell carbonate ^{14}C values using a set of values from a site on the California coast. In each case a conventional ^{14}C age is calculated by normalizing the apparent (unnormalized) ^{14}C age to $-25‰$ δ^{13}C. A reservoir correction assumed to be applicable to this coastal regime would then be applied to derive a reservoir corrected conventional ^{14}C determination. Based on contemporary shell data (Fig. 5.8), the reservoir correction for marine shell carbonates from along this coast can be inferred to be -630 years. This value would be subtracted from the conventional ^{14}C age expressions. However, there appears to be a range of several hundred years in the modern shell data from this coast. Table 5.6 presents

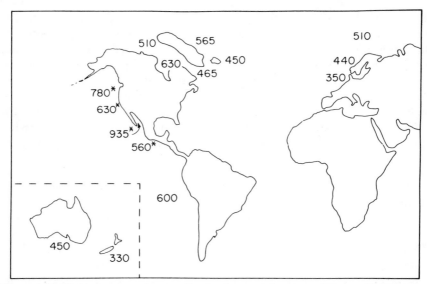

Figure 5.10   Guide to reservoir age estimates for coastal regions of the world. [Based on data presented in Robinson and Thompson (1981:48) modified on the basis of data presented in Fig. 5.9 (starred values) and in Olsson (1983b), (Cf. Mangerud and Gullikesen, 1975).]

data that allow an evaluation of the accuracy of this reservoir correction value for a portion of the California coast. In this comparison, it is assumed in each case that stratigraphic association indicates temporal equivalence. If such is indeed the case, the data would seem to confirm the suggestion from the contemporary shell data of a 200–300 year range in the reservoir correction for this coastal region. In one case, a negative reservoir cor-

TABLE 5.6
Comparison of Charcoal and Marine Shell Carbonate ^{14}C Age Determinations on Samples of Presumed Similar Age from the University Village Site, California

Lab number	Material	Unnormalized ^{14}C age (years B.P.)	$\delta^{13}C$ (‰ wrt PDB)	Conventional ^{14}C age (years B.P.)	Apparent reservoir correction (years)
UCR-957	charcoal	2610±150	−23.6	2630±150	—
UCR-958	shell	2560±150	−3.5	2900±150	−270
UCR-960	charcoal	3060±160	−25.3	3060±160	—
UCR-961	shell	3110±150	−2.9	3460±150	−400
UCR-1552	charcoal	3340±115	−24.9	3340±115	—
UCR-1553	shell	2790±130	−1.0	3170±130	+170
UCR-711B	charcoal	1320±100	−23.7	1340±100	—
UCR-711A	shell	1730±120	+0.6	2140±120	−800

rection would increase the disparity between the charcoal and marine shell values. The data presented in Fig. 5.9 and Table 5.6 illustrates the variability in initial ^{14}C activity that is exhibited by marine shell in some marine environments. In some cases, however, it appears that reservoir correction(s) are much more constant and lie within a much narrower range. For maximum accuracy, each coastal region and its subregions (e.g., estuaries and bays) would need to be evaluated independently to determine not only the general magnitude but also the degree of variability exhibited by marine shell carbonates of equivalent age. This same stricture would apply in the evaluation of reservoir effects for marine mammal remains recovered from Arctic sites. Because of the variability in reservoir effects in polar regions combined with often severe geoturbation processes observed in frozen-ground soils (Section 5.2), archaeologists concerned with circumpolar archaeology must be particularly attentive to the combinations of factors that can impinge on the accuracy and precision of ^{14}C age estimates on materials from this vast region (Arundale, 1981).

Reservoir effects are even more pronounced and can be even more variable for freshwater shell derived from lacustrine environments. Initial radiocarbon activity in such samples can be highly variable. The earliest study of this phenomenon determined that modern freshwater shell growing in a lake whose bed contained significant quantities of limestone was significantly depleted in its ^{14}C activity. Living shells in this environment exhibited apparent ^{14}C "ages" of up to 1600 years as a result of the "hard water" reservoir effect (Deevey et al., 1954). Modern gastopods from unusual environments such as artesian springs can yield seriously anomalous ^{14}C ages. There may also be the possibility of reservoir effects when dealing for example, with fish bone remains where the species of fish are bottom feeders in some marine as well as in some types of freshwater environments. However, reservoir effects in these situations should not exceed, at most, a few percent except under unusual geochemical conditions.

Terrestrial reservoir effects have also been noted in regions where volcanic gas emissions cause significant depletion in ^{14}C activity in plant materials growing adjacent to active fumarole vents. It would be expected that gases from such sources would contain CO_2 in which the ^{14}C activity would be absent due to its geological source. This magmatic CO_2 would mix with atmospheric CO_2 to yield a depressed ^{14}C activity in a highly localized area (<100-m radius) surrounding a vent or series of vents. A study of several Hawaiian localities reports apparent ages in modern plant materials of as much as 15,000–20,000 years (Chatters et al., 1969). Other studies have reported apparent ^{14}C ages on modern plant materials growing in the vicinity of volcanic gas vents ranging from about 1400 ^{14}C years, in the Eifel area of West Germany and near Thera (Santorini) in Greece, to over 4000 ^{14}C years near Monte Amiata in Italy (Saupe *et al.*, 1980; Bruns *et al.*, 1980). Ordinarily, a fumarole effect would be expected to be highly attenuated within a few kilometers of even the most active vent due to mixing with atmospheric CO_2. There apparently has been no major, worldwide variation in atmospheric ^{14}C activity attributable to vulcanism (Libby and Libby, 1972). However, in regions of major, geologically recent, volcanic activity (e.g., Iceland), wood and other terrestrial organics over relatively large areas can apparently be effected by up to several percent as a result of essentially continuous volcanic gaseous emissions (Olsson, 1979a, 1983a). Regional reservoir problems caused by fossil CO_2 injected into the atmosphere could be of concern in any area affected by major, late Quaternary vulcanism. It might also be noted that this effect should be more pronounced on short-lived samples.

Magmatic fossil CO_2 injected directly into lake waters from gas springs or fumaroles may also pose special problems in interpreting ^{14}C determinations on materials derived from such environments. In one study, modern freshwater shells from a lake with active fossil CO_2 activity exhibited apparent ages in the range of about 1000 to 2500 years (Kaufman, 1980:94). The degree to which other sample types such as bone from such environments might reflect these reservoir effects would depend on a number of factors, including the percentage of lacustrine-derived food utilized by the human or animal population. There may be a considerable range in the magnitude of the effect over relatively long periods in the same lake environment depending on changes in CO_2 injection rates, since these are often controlled by tectonic or volcanic activity. Because of this, not only should each lacustrine geochemical system be separately evaluated, but also each evaluation should include the use of multiple control samples spaced over large enough time intervals so that possible temporal variations in the reservoir effects could be identified.

Some types of terrestrial reservoir effects are difficult to evaluate on a quantitative basis since the effects are often only marginally resolvable with conventional ^{14}C analysis and it is sometimes difficult to collect a large enough suite of appropriate control samples to more precisely evaluate the suspected effects. Relatively short-lived samples growing in arid regions near to large bodies of water with depleted ^{14}C activities (e.g., Salton Sea region of southern California or the Dead Sea region of the Levant) or desert coastal areas adjacent to marine zones exhibiting significant upwelling should be especially scrutinized. Whether the "long" and "short" ^{14}C time scales that Rowe (1965) suggests are present in the corpus of ^{14}C determinations from some Peruvian archaeological sites reflect such localized conditions or can be explained by other means is not clear.

5.5.2 Calibration Procedures

The most intensively studied anomalies in ^{14}C age estimates are those caused by secular variation effects. Almost three decades of research by a number of laboratories have documented the systematic differences between "radiocarbon years" and "calendar years." Section 2.4.1 briefly reviewed the nature, magnitude, and suggested causes of these variations and their effect on deriving accurate and precise ^{14}C age estimates. Up to 1982, more than 20 publications had provided data to assist in the calibration of ^{14}C determinations in light of the ^{14}C/tree-ring data. A list of these publications appears in Klein *et al.* (1980:950, 1982:103–104). Although all of the calibration approaches take into consideration the long-term or major trend curve (Figs. 2.6 through 2.9), there has been some divergence in how the shorter-term or de Vries effects and the statistical variances associated with age estimates are to be treated. This has resulted in some variations in calibrated values depending on which calibration scheme was employed (Stuckenrath, 1977:187; Scott *et al.*, 1984). This situation has contributed to renewing concerns expressed particularly by archaeologists working in Europe and the Near East as to the overall accuracy and precision of the ^{14}C method in certain periods (e.g., Quitta, 1967; Neustupny, 1970:31; MacKie *et al.*, 1971; Mc Kerrell, 1975; Pilcher and Baillie, 1978; Freundlich and Schmidt, 1983; Waterbolk, 1983b:21–22). At the Twelfth (1985) International Radiocarbon Conference, it was proposed that calibration curves prepared by Stuiver and Pearson as published in the conference proceedings constitute the officially recommended calibration curves for the period from 2500 B.C. to the present (Mook, 1986).

The impact of calibration data on the conduct of archaeological investigations has, to date, been most far reaching in European prehistoric studies. For example, Renfrew (1973) has argued that calibrated ^{14}C values—he employed the calibration curves of Suess (1967; 1970:Fig. 16)—required what he called the "second radiocarbon revolution": a major revision in traditional understandings of the factors involved in the origins of several important archaeological features of "European barbarism," the megalithic chamber tombs, the temples on Malta, European metallurgy, and Stonehenge in England. The generally accepted view at that time saw the first three of these developments as having their ultimate origins in the Near East, while Stonehenge reflected the inspiration of Mycenaean Greece. European advances in areas such as monumental architecture and metallurgy were initiated or stimulated by diffusion of knowledge initially from Egypt and southern Mesopotamia through the eastern Mediterranean and Aegean and hence to Europe. Renfrew (1973) argued that diffusionist views were seriously undermined by the calibrated ^{14}C values, which placed the megalithic tomb structures earlier than the Egyptian pyramids, the Malta temples before any Near Eastern counterparts, and copper metallurgy in the Balkans earlier than in Greece. Also, Stonehenge was completed before the Mycenean civilization in Greece began.

The need to calibrate ^{14}C values, i.e., to adjust "radiocarbon years" to "calendar years," arises in situations in which ^{14}C-based chronologies are to be compared with those based on "real" (i.e., siderial or calendar) time. As we have noted, calibrated ^{14}C values have been important in understanding temporal relationships in prehistoric Europe due to the fact that much of the traditional chronological structure is either directly or indirectly linked into the historical- and calendar-based chronologies of the Near East. Important insights into the interpretation of the history of construction of several types of European Medieval wooden structures have also been gained with the recognition of the need to interpret ^{14}C values in light of the de Vries effects during the later Middle Ages (Horn, 1970; Berger, 1970b; Fletcher, 1970). Calibration would also be required if one wished to compare ^{14}C and dendrochronologically based age estimates in such regions as the southwestern United States or where calendar systems were employed, such as in the Maya region of Mesoamerica. In theory, such procedures would also be required if ^{14}C values were to be compared with age estimates derived from other physical dating methods, e.g., obsidian hydration or amino acid racemization. However, in view of the fact that rate structures for these techniques were typically estimated on the basis of ^{14}C data and since the precision available from these techniques is much less than that available with ^{14}C values, only rarely would any benefit be derived from calibration procedures (cf. Mook, 1983b).

In Section 3.3.1, we noted the use of ^{14}C determinations in discussions concerning the Maya correlation formulas used to translate Maya calendar dates into their Western calendar equivalents. Following the identification of the problems of presample growth, a large series of ^{14}C values were obtained on Maya-dated wooden lintels. These data became available just as the identification of the secular variation/de Vries effects were becoming widely recognized. The degree of calibration needed to bring ^{14}C values into alignment with "calendar years" was accomplished by the use of what was called the "effective half-life." Ralph had determined that an increase in the half-life figure of about 3% would bring ^{14}C determinations into line with known-age materials. By this means, the ^{14}C determinations on the Maya lintels were thus indirectly "calibrated."

Klein *et al.* (1982) has provided a convenient summary of calibration data from five ^{14}C university laboratories [Arizona, Groningen, University of California, San Diego (La Jolla), Pennsylvania, and Yale] for the last

7200 ^{14}C years. This is the equivalent in calendar time of about 8000 years or about 80% of the Holocene period. These ^{14}C/dendrochronological data have been presented in two formats. The first is as plots of the relationships between ^{14}C values (expressed in ^{14}C years B.P. with $t_{1/2} = 5568$ years) and equivalent dendrochronological ages (expressed in A.D./B.C. calendar years). The second format lists in tabular form the relationship between a single ^{14}C age value rounded to the nearest decade and the equivalent calibrated value expressed as an age range at a statistical confidence level of 95% (2σ) as a function of the 1σ measurement uncertainty of the ^{14}C value for which calibration data are desired. Figure 5.11 illustrates a segment of the plotted data for the early third and fourth millennia B.C., and Table 5.7 lists an example of the format of the calibration tables for the period from 5500 to 5600 ^{14}C years B.P. (Klein *et al.*, 1982:128–129). Note in Fig. 5.11 that associated with each time interval are four time bands

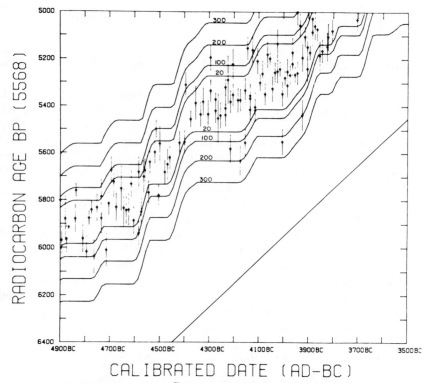

Figure 5.11   Calibration data for the ^{14}C time scale, 5000 to 5800 ^{14}C years B.P. [From Klein *et al.* (1982:116).]

TABLE 5.7
Calibrated Radiocarbon Age Ranges at 95% Confidence Interval for Different Statistical Variance in Range of 5400 to 5500 ^{14}C years B.P.[a]

Radiocarbon age (B.P.) with 5568 half-life	1 σ range of individual ^{14}C value			
	± 50 Years B.C.	±100 Years B.C.	±150 Years B.C.	±200 Years B.C.
5500	3995–4530	3960–4545	3915–4565	3890–4715
5490	3980–4520	3950–4545	3910–4650	3885–4715
5480	3975–4440	3945–4540	3905–4560	3885–4705
5470	3965–4435	3935–4540	3900–4555	3880–4700
5460	3960–4435	3930–4535	3900–4555	3880–4695
5450	3955–4430	3920–4530	3895–4550	3875–4690
5440	3945–4430	3915–4525	3890–4550	3870–4680
5430	3940–4425	3910–4520	3890–4545	3870–4565
5420	3930–4425	3905–4440	3885–4545	3865–4565
5410	3925–4420	3900–4435	3885–4540	3860–4560
5400	3915–4420	3895–4435	3880–4540	3860–4560

[a]From Klein et al. (1982:128–129).

defining the calibration ranges for different 1σ statistical ranges (±20 to ±300 years) of a ^{14}C value to be calibrated.

In tabular form for the periods in excess of 1000 ^{14}C years B.P., each decade value carries with it six calibrated age ranges at the 95% confidence level. Table 5.7 illustrates four calibrated age ranges for ±50 to ±200 years. As discussed in Section 2.2 and as illustrated in Fig. 2.12, calibration ranges will vary as a function of the statistical precision of the calibration values, the defined shape of the calibration curve at various points, as well as the statistical variance carried by the ^{14}C value to be calibrated. At 5410 ^{14}C years B.P., for example, the calibrated age range for a value with a 1σ counting error of ±50 years (this can be used for counting errors of ±35 to ±75 years) is 4420–3925 calendar years B.C. or an age range of 495 years. The calibrated age range listed under the ±100 column (for counting errors of ±75 to ±125 years) is 4435–3900 calendar years B.C. for a 2σ calibration range of 535 years. The difference between the ±50 and ±100 calibration ranges at this point in the calibration curve is 40 years. However, for a ^{14}C value of 5450, the difference between the ±50 and ±100 calibration ranges is 135 years.

5.6 EVALUATING RADIOCARBON AGE ESTIMATES

Our discussion has noted four factors that can influence the accuracy and precision of ^{14}C age estimates: (i) sample provenance, (ii) sample composition, (iii) experimental factors, and (iv) systemic factors. We have suggested that the primary responsibility of the archaeologist is to clearly delineate the nature of the behavioral event, cultural phenomenon, or object for which temporal placement is being sought and then, if necessary in collaboration with a geologist and/or other Quaternary environmental specialist, identify the nature of the relationship or association between the event/phenomenon and sample materials to be used for the ^{14}C measurement. The collection of samples in the field should be accompanied by detailed observations concerning possible sources of contamination and a view that chronological judgments can most effectively be determined by employing suites of ^{14}C determinations on multiple samples drawn from the same context or with multiple ^{14}C determinations obtained on different fractions of the same sample. The use of a single ^{14}C sample to assign age to a feature or stratigraphic level should be avoided.

The submission of samples to a laboratory should follow consultation with the laboratory director concerning the various factors that could affect the accuracy or precision of the particular samples to be used. A review of sample provenance, types of samples collected, amount of sample available for analysis, possible contamination factors, counting precision required, possible variations in ^{13}C values, and nature of reservoir corrections are examples of topics that might be addressed. The conventions of the laboratory with respect to how statistical errors are calculated, how reservoir corrections are made, or different methods of sample pretreatment might also be discussed.

Unless a laboratory deviates from the standard format, ^{14}C determinations will be expressed in the conventional form as discussed in Sections 2.1 and 4.6. As we have noted, a conventional radiocarbon age assumes the following: (i) the use of 5568 (5570) years as the ^{14}C half-life, (ii) the use of A.D. 1950 as 0 B.P., (iii) the direct or indirect use of NBS oxalic acid as a standard, (iv) normalization of all sample activities to a base of $\delta^{13}C = -25‰$ with respect to PDB, and (v) the assumption of the constancy of atmospheric ^{14}C levels over the ^{14}C time scale. If these conventions have been followed, marine shell carbonates will already have had their ^{14}C values adjusted in light of their measured or estimated $\delta^{13}C$ values. It will be necessary to apply a reservoir correction to these values. The amount will be variable depending on the region from which the samples were derived. These corrections need to be discussed with the laboratory director from which the ^{14}C values were obtained to insure appropriate application of the correction values.

Following the application of the appropriate corrections (if such are required), *individual ^{14}C age estimates* may be evaluated in light of their statistical values and secular variation (major trend/de Vries) effects. One suggested approach is set forth below:

1. Convert the conventional ^{14}C age expression into its equivalent age range. (For recognized high-precision laboratories, if the standard error

is less than ±40 years, increase it to that value. For ^{14}C determinations from conventional laboratories, if the standard error is less than ±80 years, increase it to that value.) Double the error expression and use this value in determining the age range at the 95% confidence level. If appropriate, convert the age range values into their *apparent* A.D./B.C. equivalents to facilitate comparisons with calibrated values. Note the magnitude of the age range.

> Example: 3550±90 ^{14}C years B.P.
> 3370–3730 ^{14}C years B.P. (95% confidence level)
> 1420–1780 ^{14}C years B.P. ($-$ 1950 expression)
> Age range = 360 years

2. Examine calibration data associated with the ^{14}C value to determine if there are any significant de Vries effects to consider. *Use the original standard deviation assigned to the cited ^{14}C value.* Note the magnitude of the age range of the calibrated value.

> Example: 3550±90 ^{14}C years B.P.
> 1680–2180 calendar years B.P. (95% confidence level)
> (after Klein *et al.*, 1982:134)
> Age range = 500 years

3. Utilize the age range of the calibrated value to interpret the ^{14}C B.P. age range. Cite sample number, conventional ^{14}C age (^{14}C years B.P.), δ ^{13}C value, most likely age range at 95% confidence interval (in ^{14}C years B.P.), [*if appropriate*, calibrated age interval (calendar years B.C./A.D.) with the calibration approach identified], and publication source.

> Example: UCR-1692, 3550±90 ^{14}C years B.P., δ^{13}C = -24.68 ‰
> PDB, ca. 3300–3800 ^{14}C years B.P., 95% confidence
> interval [1680–2180 calendar years B.C. based on
> Klein *et al.*, (1982:134)].

The calibrated age interval in calendar years B.C./A.D. has been placed in brackets to emphasize that the citation of calibrated ^{14}C age expressions is most relevant in situations where the ^{14}C time scale is being compared with calendar or sidereal time. *Where the ^{14}C scale represents the only basis on which time placement chronological inferences can be made, it would seem reasonable that temporal comparisons and summaries be simply expressed in radiocarbon time.*

However, awareness of the calibrated values where these are expressed in 95% confidence intervals is important in assessing the overall precision that can be attached to any single ^{14}C age estimate. Figure 5.12 illustrates the usefulness of such an approach by representing a series of ^{14}C deter-

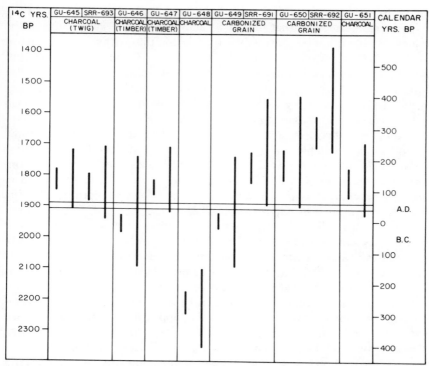

Figure 5.12   Radiocarbon determinations on samples associated with the Cadbury massacre. For each ^{14}C determination, two values have been plotted. The first is the conventional ^{14}C age. The second is the calibrated value expressed at the 95% confidence interval following the approach of Klein *et al.* (1982). [Adapted from Table 2 in Campbell *et al.* (1979).]

minations obtained on samples associated *with a single event* recovered during excavations carried out at Cadbury Castle in England. All but one of the ten ^{14}C measurements were obtained on samples of charcoal and carbonized grain recovered from a thin (50–150 mm) burned layer contained in the remains of a stone guard chamber dating to the late Iron age—destroyed during a Roman military campaign that resulted in the massacre of the defenders of the garrison. Because of this, the event is referred to as the "Cadbury massacre." The tenth sample was collected from the adjacent portal and appears to be associated with the burning of the gates. The Cadbury massacre and the conflagration that followed have been dated on the basis of historic criteria to between A.D. 45 and A.D. 61 (Campbell *et al.*, 1979).

An initial examination of this suite of values using only the age equivalents associated with the average or midpoint of the conventional 1σ ^{14}C age expressions, as would, for example, be represented in a table of ^{14}C values, might conclude that there was an unacceptably large range in the values. Would it not be reasonable to conclude that with a range of about 1650 to 2200 ^{14}C years B.P., one or more of the ^{14}C dates should be considered anomalous and thus discarded from consideration? In one case, duplicate samples analyzed by two laboratories (GU-649/SRR-691) differ by about 170 years. A representation of the Cadbury Castle ^{14}C dates such as that in Fig. 5.12 makes it easier to appreciate both statistical and calibration parameters for each sample. For each sample, the conventional ^{14}C values are first plotted using their 1σ standard deviations (Sections 1.3 and 4.7). Then, calibrated ^{14}C values are represented as bands of uncertainty representing the 95% (2σ) confidence interval as evaluated in light of the characteristics of calibration data for that period following procedures outlined in Klein *et al.* (1982).

In each case, the calibrated values encompass the conventional age expressions since for this interval radiocarbon time does not significantly deviate from calendar time. De Vries effects are indicated by the varying lengths of the calibration intervals. For eight of the ten values, the 2σ ranges of the conventional age expressions as well as the calibrated values overlap with the presumed known-age of the samples. However, the remaining two samples exhibit ^{14}C values that appear to be anomolous. Since beams were used in the construction of the guard chamber, the presample growth error could be invoked as one means of explaining GU-648. SRR-692 was obtained on a short-lived type of sample and appears to be several centuries too young. However, if both ^{14}C values (GU-650 and SRR-692) are combined, the calibration interval associated with the average value overlaps with the assumed age of the samples. If it is assumed for statistical purposes (e.g., Long and Rippeteau, 1974) that each ^{14}C value represents an independent observation of a single event, then an average value of 1890 ± 10 ^{14}C years for the Cadbury massacre could be computed. Using the 5730 half-life, the value becomes 1940 ± 10.

Manipulating the Cadbury ^{14}C values in such a manner may be justified in statistical terms, but it also may obscure the value of this type of data set to illustrate an important consideration in the evaluation of ^{14}C evidence. As has been emphasized previously, standard ^{14}C age estimates are inherently time segments of varying magnitude. The range of primary inherent uncertainty that must be assigned to ^{14}C age expression relates most directly to the magnitude of the inherent precision of the ^{14}C measurement as well as the characteristics of the various de Vries effects for the time period involved (Scott *et al.*, 1983; Stenhouse and Baxter, 1983). Added to these primary considerations would be variability introduced as a result of adjustments in other geophysical/geochemical parameters reviewed in this volume. If ^{14}C values from a number of different laboratories are being compared, an added consideration would be a realistic consideration of interlaboratory variability (Baxter, 1983). Except in unusual circumstances, the typical range of *overall minimum uncertainty* of an *individual* ^{14}C age determination for middle and late Holocene materials would be at least 200 years. For the early Holocene, the minimum overall band of uncertainty would typically be at least 300 years. It can be legitimately argued that the distribution of ^{14}C values exhibited by the materials from the Cadbury massacre are not excessive or unusual. They represent a realistic portrayal of the overall variation in ^{14}C values that might be expected from an archaeological locality (cf. Waterbolk, 1983a:63). A similar range in ^{14}C values obtained on a series of samples dating to the brief Norse occupation at L'Anse aux Meadows, Newfoundland, illustrates this same point, although in this case the possible use of driftwood creates additional interpretive problems (Nydal, 1983:110–113, 116).

The Cadbury ^{14}C values and similar, relatively large ^{14}C data sets from presumed "single event" contexts (e.g., Burleigh and Hewson, 1979) provide a cautionary note and commentary on the inherent limitations of the ^{14}C method to distinguish temporal increments in units of less than 2–3 centuries at reasonable levels of precision except under special circumstances. Attempts to employ ^{14}C determinations without appropriate regard for the full range in inherent and effective variability have sometimes led to unrealistic expectations. For example, this has resulted in a questioning by some archaeologists of the general validity of the calibrated radiocarbon time scale as documented by the bristlecone pine data as compared to historically based chronologies. This uncertainty or mistrust was exacerbated by the variety of calibration schemes that have been put forth. This has been the case for the Near East and in particulary with respect to the Egyptian chronology. In the Egyptian case at least, it appears that where secure contexts have been documented for samples and appropriate attention has been given to the relevant geophysical/geochemical parameters, there has generally been reasonable agreement between appropriately corrected and calibrated ^{14}C age determinations and age estimates based on historical/archaeological criteria (e.g., McKerrell, 1975; Mellaart, 1979; cf. Save-Soderbergh and Olsson, 1970; Clark and Renfrew, 1972; Clark, 1978; Weinstein, 1980; Kemp, 1980).

Under special circumstances, it is sometimes possible to derive unusually precise ^{14}C age estimates on wood or charcoal samples if they are composed of a sufficient number of tree ring segments that lie within the time span documented by the calibration data (Fergusson *et al.*, 1966; Suess and Strahm, 1970; cf. Beer *et al.*, 1979; Kruse *et al.*, 1980). The procedure involves the identification in the unknown age samples of a de Vries effect pattern of short-term ^{14}C variations. This can be accomplished by obtaining ^{14}C measurements on an appropriate number of individual tree rings in the sample. There can then be an attempt to match the de Vries effect variations or "wiggles" exhibited by the unknown-age sample set against the known-age short-term variations of a high-precision calibration curve. This procedure, dubbed "wiggle matching," requires that the amount of sample available from each ring segment be sufficient so that sample standard deviations in the ^{14}C analysis of the samples being dated be comparable to that of the ^{14}C values comprising the calibration curve itself (i.e., 1σ statistical variance of ± 15–30 years) and that the number of tree ring segments be of sufficient number to achieve statistically significant matches (Clark and Renfrew, 1972; Clark and Sowray, 1973). Unfortunately, samples that would permit this type of analysis are rarely available from most archaeological contexts.

Although the emphasis in this section has been on evaluation, correction, and calibration of individual ^{14}C determinations, it would be appropriate to emphasize again that attention should be focused on the general *pattern* of ^{14}C values rather than on any individual ^{14}C age estimate. Only with a *suite* of ^{14}C values can problematic results be critically identified (Walker *et al.*, 1983:435). The existence of apparently discordant ^{14}C data should be the stimulus for additional studies to determine, if possible, the reasons for anomalous values. In analytical terms, there are probably a relatively small number of "bad" ^{14}C values (i.e., involving major laboratory error in the measurement of the ^{14}C activity of a sample.) There are, however, a number of ^{14}C determinations for which there is no immediate explanation for the apparently seriously disconcordant values. A serious disconcordance may be defined as a situation in which a suite of ^{14}C values appropriatedly pretreated, normalized, reservoir-corrected, and interpreted in light of secular variation effects are offset consistently by more than 500 ^{14}C years from an expected age assignment. In most cases, the resolution of the problem requires a reexamination of the context from which the samples were obtained. In other cases, the analysis of additional samples, either those which are duplicates of the original sample material or material collected from an identical archaeological/geological environment, would be required. It might be noted in this context that ^{14}C data should not stand alone to support chronological inferences. Each ^{14}C value should be viewed within the context of a specific issue or problem. The results of the ^{14}C analyses should be consistent with relevant stratigraphic, paleoenvironmental, paleontological, historic, and/or archaeological inferences. If it is not, investigations need to continue until the anomaly is resolved.

5.7 IMPLICATIONS OF RADIOCARBON DATING IN ARCHAEOLOGICAL STUDIES

The most immediate and obvious impact of the ^{14}C method on the conduct of archaeological research has been the availability of a fixed-rate temporal scale of worldwide applicability. In providing a common frame of temporal reference for the late Pleistocene and Holocene, ^{14}C data made a *world* prehistory possible by contributing a time scale that transcends local, regional, and continental boundaries. As Grahame Clark (1970:38) noted, radiocarbon "has contributed more than any other single factor to complete the world coverage of prehistoric archaeology, not to mention the way it has helped scholars to synchronize phenomena in different parts of the world." Without the ^{14}C time scale, prehistorians would still be, in the words of J. Desmond Clark (1979:7), "foundering in a sea of imprecisions sometimes bred of inspired guesswork but more often of imaginative speculation." It can be legitimately argued that temporal intercomparability was as important a characteristic of ^{14}C values as the exact degree of accuracy or precision. Fortunately, the temporal framework itself turned out to be amazingly accurate given the number of assumptions that had to hold to rather narrow ranges. This is the context of Renfrew's "first radiocarbon revolution" (Renfrew, 1973:48–68). The geologically late beginning of the postglacial period and the surprising antiquity of agriculture and sedentary village societies in Southwestern Asia as well as several areas of Western Europe and Mesoamerica are examples of major unanticipated results of the first two decades of ^{14}C dating.

In addition to providing the potential for a time scale employing chronometric units which, when appropriately corrected and calibrated, could be directly compared with historic chronologies, ^{14}C data can play other important functions in the conduct of archaeological investigations. It has been suggested, for example, that the advent of ^{14}C dating led to a noticeable improvement in archaeological field methods (Johnson, 1965:764). The purpose of these more refined methods of data recording was to determine more precisely the association of samples with archaeological levels. In some cases, the initial motivation to improve was to demonstrate that a ^{14}C age estimate was erroneous. In most cases, the essential correctness of the ^{14}C value was sustained. However, the effect on field methods was permanent. In addition, Waterbolk (1983a) has noted that ^{14}C values can provide a valuable independent means of confirming the identification of archaeological features where stratigraphic or typological data is incomplete or problematical. Radiocarbon data can also be employed to provide an independent correlation of sequences built up on the basis of strictly archaeological criteria with environmental sequences derived from geomorphological data or from the study of major regional or subcontinental floral and faunal successions. Also, ^{14}C values can be very useful in guiding lengthy or multiseason archaeological field studies particularly in the case of complex or disturbed stratigraphic contexts.

A less immediately recognized contribution of ^{14}C data has been the fact that ^{14}C-based age estimates provide a means of deriving chronological relationships independently of assumptions about cultural processes and totally unrelated to any type of manipulation of strictly archaeological materials (cf. Willey and Phillips, 1958:44; Taylor, 1978:63; Dean, 1978:226). When pressure to derive chronology primarily from the analyses of artifact data was released, inferences about the evolution of human behavior based on variations in environmental, ecological, or technological factors could be aggressively pursued by employing an independent chronological framework (cf. Clark, 1984). In the United States, the rise of the "new archaeology" in the 1970s took place in this context. Lewis Binford has reflected that ^{14}C chronology "has certainly changed the activities of archaeologists, so that now, in many ways for the first time, they direct their methodological investments toward theory building rather than towards chronology building" (quoted in Gittens, 1984:238).

This is illustrated by Renfrew's "second radiocarbon revolution" in that calibrated ^{14}C values required a reevaluation of traditional views of the cause of the introduction of important innovations in prehistoric European culture. In place of diffusionist explanations for the new elements, Renfrew emphasized local factors such as increasing population pressures, trade and exchange systems, as well as the manner in which social organization interrelated with developing indigenous economies and technologies (Renfrew, 1973:248). As far as the study of British prehistory was concerned, Atkinson (1975:174) characterized the impact of both conventional and calibrated ^{14}C data as "radical . . . therapy" for the "progressive disease of 'invasionism'." For the future, the relatively large number of ^{14}C values for a number of areas of the world can provide a stimulus for a periodic critical appraisal of the contribution of ^{14}C data to an understanding of cultural processes that have operated in a given region. This appraisal could also identify those archaeological problems and issues for which carefully designed future research, including, if appropriate, additional ^{14}C analyses, could provide important additional insights into the causes of the evolution of human behavior in a given region. In fact, temporal frameworks used in association with these types of investigations require even finer control over time than has been traditionally required. Studies dealing with the evolution of, for example, settlement systems or exchange networks require that it be possible to determine which sites in a region were occupied contemporaneously (Read, 1979). Only with the most carefully crafted archaeological research design that focuses in detail on sample collection strategies, with close attention to sample/context parameters, as well as addressing all of the relevant geochemical/geophysical factors discussed in this volume, could the required accuracy and precision required in the ^{14}C data be approached.

The great technical success of the ^{14}C method has also provided a major impetus for interdisciplinary or contextual (Butzer, 1978) studies in archaeology (cf. Willey and Sabloff, 1980:155). Stimulated by European efforts, this tradition was already underway in American archaeology in the immediate pre-World War II period as exemplified in the research conducted as part of the excavations at the Boylston Street Fishweir excavated in Boston, Massachusetts (Johnson, 1942, 1949). However, it was the ^{14}C dating method that provided one of the major catalysts that moved archaeologists increasingly to direct their attention to analytical and statistical approaches in the manipulation and evaluation of archaeological data (cf. Thomas, 1978:323) as well as to maintain long-term collaborative relationships with colleagues in other scientific disciplines (cf. Barnard, 1982). It was therefore entirely congruent that an early proponent of interdisciplinary studies in archaeology, Frederick Johnson, also became an early advocate of the ^{14}C method among his colleagues. The general theme of his concerns at the inception of the method is still relevant as we look forward to the continuing expanding horizons of the application of ^{14}C in archaeological studies:

> [P]rogress in the development of . . . [radiocarbon dating] depends to a large degree upon the character of the collaboration [between archaeologists and other scientists]. The laboratory procedure involves theories in physics and chemistry which for the most part are outside the experience of almost everyone who has a sample to be dated. On the other hand, the results secured are of little consequence unless they are directly or indirectly related to some stratigraphic sequence. The value of the laboratory results is enhanced by critical evaluation by other scientists. Most particularly, the reverse is true. This involves continual examination of all basic theory and hypotheses by everyone concerned. The future value and usefulness of the method depends in large measure upon the success of continued collaboration between physicists, archaeologists, geologists, botanists, and others [Johnson *et al.*, 1951:62].

This injunction has assumed a new significance as a result of the recent introduction of AMS to accomplish ion or direct counting of ^{14}C, ushering in the third generation of ^{14}C studies. As we have noted (Section 4.5), AMS methods permit measurements to be conducted on milligram amounts of sample materials and open up the possibility of extending the ^{14}C dating range beyond the current boundary of between 40,000 and 50,000 years for samples collected from typical archaeological contexts. For the archaeologists, the increasing utilization of milligram-size samples will require an even more rigorous attention to the evaluation of geological, geochemical, and archaeological contexts of samples. The need for interdisciplinary cooperation and collaboration has become even more critical as AMS technology assumes an ever increasing role in the ^{14}C analysis of archaeological and other late Quaternary paleoecological materials over the next decade and beyond.

CHAPTER 6

RADIOCARBON DATING IN HISTORICAL PERSPECTIVE

6.1 DISCOVERY OF RADIOCARBON

In his retrospective reflections, or "ruminations" as he once characterized them (Libby, 1970a, 1970b, 1982) on the roots of the ^{14}C method, Libby was careful to emphasize his indebtedness to several earlier investigators. However, his recall of his own discovery process contained several versions and interpretations. This is understandable since during the period of the early ^{14}C studies at Chicago he was involved in many different research topics, as a review of his publications for this period will show (Berger and Libby, L.M., 1981). Thus, in a reconstruction of processes involved in the development of the ^{14}C method, the recall of his co-workers and collaborators is invaluable and indispensible. Figure 6.1 lists the major concepts and discoveries instrumental to the process by which the ^{14}C technique came to be developed.

147

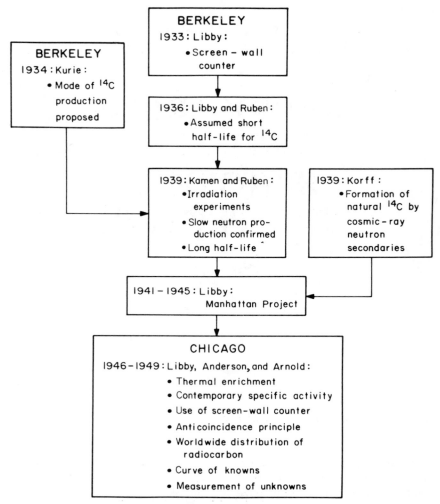

Figure 6.1 Major events and researchers involved in the development of radiocarbon dating.
[After Taylor (1978:3).]

Radiocarbon is one of a number of isotopes that had been produced artificially in the laboratory before being detected in the natural environment. Both Martin Kamen (1963:584) and Libby (1967:3) noted that the first published suggestion of the existence of ^{14}C was contained in a 1934 paper by Franz Kurie as a somewhat unlikely interpretation of the effect of neutron bombardment on nitrogen in a cloud chamber (Kurie, 1934). Since the Berkeley cyclotron was used in the experiments, Libby would later emphasize that ^{14}C was a "Berkeley discovery" (Libby, 1964a:1). Kurie's paper appeared the year following Libby's receipt of his doctoral degree in chemistry at Berkeley and his appointment there as an instructor. Libby's 26-page dissertation was entitled "Radioactivity of Ordinary Elements, especially Samarium and Neodymium: Method of Detection" filed in the spring of 1933 (Libby, 1933).

While still an undergraduate, Libby had built the first Geiger–Müller (GM) type radiation detector assembled in the United States (Libby, 1964b:8; Libby, L. M., 1981, n.d.). The GM tube design had been first published in 1928 in a German scientific periodical by Hans Geiger and Walther Müller. Libby's original GM-type detector consisted of a brass tube down the length of which was stretched an iron wire. It used air at a reduced pressure as the counting gas. A microammeter connected to the grid element of a two-stage amplifier circuit detected the presence of ionizing radiation. The description of this amplifier constituted Libby's first official publication (Libby, 1932).

In his dissertation research, Libby's counter envelope was made of glass. A wire grid or "screen wall" was substituted as the cathode of the counter instead of using a solid metal wall. In this way, very weak radioactivities in solid samples could be detected. His samples of rare earth elements were mounted on the inside surface of a glass sleeve that could be moved in and out of the counter's sensitive region on glass beads acting as ball bearings (Libby, 1933, 1934; Libby and Lee, 1939). This design was the forerunner of the type of screen-wall counter that Libby would employ more than a decade later in all of his ^{14}C experiments with solid carbon counting at the University of Chicago.

In the mid-1930s, the soon-to-be Nobel laureate, E. O. Lawrence (Physics prize in 1939 for the development of the cyclotron and research in artificial radioactivity) organized the Radiation Laboratory at Berkeley (now the Lawrence Berkeley Laboratory). In 1936, ^{14}C recoil tracks were used to calibrate cloud chamber experiments, but little was known about the isotope's physical characteristics. Following Kurie's suggestion of ^{14}C production with neutrons, Samuel Ruben, a Berkeley chemistry graduate student, was set to work to produce ^{14}C by irradiating ammonium nitrate in Lawrence's then newly completed 27-inch cyclotron. At that time, ^{14}C was assumed to have a half-life of, at most, a few months. The experiment, undertaken on the assumption of a short half-life, failed to produce any measurable amounts of ^{14}C. Ruben went on to complete his degree by pursuing other topics. No determined effort to continue work on ^{14}C was immediately undertaken in view of the presumed short half-life. This was because the work of the Lawrence laboratory between 1937 and 1939 was almost entirely focused on the production of isotopes for medical and biological research. Lawrence believed that this promised the best chance of long-term support for the operation and expansion of cyclotron research at Berkeley (Kamen, 1963:586; Libby, 1967:4; Libby, 1979a).

In late 1939, a basic policy question was raised concerning the practical worth of radioisotopes in biomedical research. Lawrence now ordered a maximum effort to determine once and for all whether long-lived radio-active isotopes existed for any of the biologically important elements—especially hydrogen, nitrogen, oxygen, and carbon (Kamen, 1963:588). The work on carbon, undertaken by Martin Kamen and Ruben, began by exposing graphite to deuteron bombardment in the 37-inch cyclotron (a deuteron is the nucleus of deuterium, an isotope of hydrogen; deuterium oxide is "heavy water"). Following the bombardment, the graphite was burned to CO_2, precipitated out as $CaCO_3$, and used to coat the inside of Libby's "screen wall" counter. The counter registered the presence of artificial ^{14}C and the surprised researchers came to realize that the half-life had to be orders of magnitude in excess of that previously assumed.

All of the initial experiments at Berkeley had been based on the assumption that the reaction of deuterons with ^{13}C rather than neutrons with ^{14}N would be the *most likely* reaction in the formation of ^{14}C. To exclude the possibility of a favored reaction of neutrons with ^{14}N, an experiment was devised in which a solution of saturated ammonium nitrate was irradiated with neutrons. The expectation was that no detectable amount of ^{14}C would be produced. Instead, a relatively small amount of precipitate paralyzed the screen wall counter because of the high ^{14}C count rate. Within a few months of February 1940, despite strong theoretical arguments to the contrary, Kamen and Ruben experimentally demonstrated that the thermal ("slow") neutron mode of ^{14}C production was heavily favored and that the half-life was in excess of 1000 years, somewhere in the range of 10^3 to 10^5 years (Ruben and Kamen, 1941; cf. Kamen, 1985:122–146). (Tragically, Ruben's promising scientific career was cut short in an accident at Berkeley during World War II.)

Sometime in the mid-1930s, Libby had become aware of the work of Serge A. Korff. A cosmic ray physicist at the Bartol Foundation in Pennsylvania and later at New York University, he had been measuring the increase of cosmic rays with increasing altitude in the atmosphere by sending Geiger counters aloft in balloons. Korff had apparently become interested in determining if there were neutrons in cosmic rays as a result of a statement by the Nobel physicist Robert A. Milliken that there were probably none (L. M. Libby, n.d.). Neutrons—neutral elementary particles—had by this time been produced in the laboratory, but it was not known if they existed in natural radiation received by the earth. Korff set to work to build a counter that would be sensitive only to neutrons. Unknown to Korff, Libby had also undertaken to develop a neutron counter using boron trifluoride as a counting gas. Based on information indirectly obtained from a colleague of Libby's at Berkeley, Korff used boron trifluoride to which he added argon as his counting gas. By sending his counters aloft in balloons, Korff found an increase of neutrons with altitude up to about 16 km, after which there was a rapid decrease. These data were interpreted to indicate that neutrons existed naturally but were secondary radiation, a product of the collision of high-energy cosmic rays with nuclei of the gaseous components of the earth's atmosphere. The decrease at the top of the atmosphere was attributed to the escape into space of some of the neutrons so formed (Libby, 1955:1). Korff noted that the neutrons that remained would be slowed down by collisions with atmospheric gas nuclei and that these "slow" or thermal neutrons would disappear by being captured by nitrogen forming ^{14}C. This was apparently the first published prediction that ^{14}C would exist *in nature* (Korff and Danforth, 1939; Montgomery and Montgomery, 1939; Korff, 1940:134; Libby, L. M., n.d.). Libby later stated that the origin of ^{14}C dating involved his reading of the Korff and Danforth (1939) publication, which reported finding neutrons in the atmosphere: "As soon as I read Korff's paper . . . that's carbon dating" (Libby, 1979a, pp. 33, 40).

The entrance of the United States into World War II quickly redirected the pursuit of nuclear studies. Many of those affiliated with Lawrence at Berkeley became attached to one of the Manhattan Project centers. Libby, on his first sabbatical leave from Berkeley, interrupted his tenure as a Guggenheim Fellow at Princeton in August 1940 and volunteered his services to the Manhattan group at Columbia University headed by Harold Urey (1934 Nobel Prize in chemistry for the discovery of deuterium). During the war years, despite his preoccupation with the principal matter at hand (developing a thermal diffusion method that would work on a large scale to separate ^{235}U from ^{238}U for use in a nuclear weapon), Libby considered a number of "useless and impractical" topics including ^{14}C chemistry (Libby, 1979a:30–31). One obvious issue was the question of the half-life of ^{14}C. Attempts undertaken during the Manhattan Project years resulted in values of 26,000 ± 13,000 and 21,000 ± 4,000 years (Libby, 1952a:34, footnotes 2 and 3). Only with the immediate postwar use of more precise mass spectrometric data to determine more precisely the isotopic composition of the samples used in the experiments and the large amounts of ^{14}C that could be produced artifically would a more accurate value be measured.

6.2 LIBBY AT CHICAGO

At the conclusion of World War II, Berkeley had promoted Libby to a tenured position as Associate Professor. However, looking forward to continuing his relationship with Harold Urey, Libby accepted an appointment as Professor of Chemistry in the Department of Chemistry and Institute for Nuclear Studies (now Enrico Fermi Institute for Nuclear Studies) at the University of Chicago. At 36 years of age Libby was, at that time, the youngest full professor at Chicago. He arrived to take up his position in October 1945 with a number of research projects in mind. He later insisted that he originally believed the "notion of radiocarbon dating" to be "beyond reasonable credence" (Libby, 1980:1018–1019). He later recalled that he initially decided to pursue the ^{14}C project in secret: "[he] did not tell anyone of his final goal of proving radiocarbon dating would be able to reveal the history of civilization because he felt that if he talked about such a crazy idea he would be labeled a crackpot and would not be able to get money to fund his research nor students to help him" (L. M. Libby, n.d.). When later asked to identify the most difficult and critical part of the development of the technique, he responded "Being smart enough to keep it secret until it was in hand. . . . I don't care who you are. You couldn't get anybody to support it. It's obviously too crazy." Apparently only Harold Urey, Libby's mentor at Chicago, initially knew of the goal of the research (Libby, 1979a:35, 45).

The first published hint of the direction in this thinking appeared in a short paper in *Physical Review* for June 1946. By the time this note appeared, Libby had been informed of the results of a much more accurate measurement of the ^{14}C half-life at about 5000 years. In his *Physical Review* paper, he noted that if the half-life of ^{14}C was much greater than 1000 years, then a balance between ^{14}C production and decay would exist in living organics, and he predicted what the specific ^{14}C activity of modern carbon would be. Most importantly, however, he predicted that there would be a significant difference in the ^{14}C activity between biological and fossil carbon (Libby, 1946). The kernel of the conception of ^{14}C dating is contained in this paper. It is interesting that this paper also contained what apparently is the earliest use by Libby of the term "radiocarbon," a contraction of "radioactive carbon." There is a possibility that Libby himself was the first to employ this term for ^{14}C (F. Johnson, personal communication) paralleling a common prewar usage for radioactive isotopes, e.g., "radiocopper," "radiocobalt," and "radiocadmium" (E. C. Anderson, personal communication). In 1934, for example, Aristide V. (von) Grosse, who would be one of Libby's collaborators on his first major

^{14}C experiment, had forecast the existence of "cosmic radio-elements" as one effect of cosmic rays impinging on the earth's surface (Grosse, 1934). (Strictly speaking, of course, radiocarbon is an imprecise term since there are other known radioactive isotopes of carbon [^{10}C, ^{11}C, ^{15}C] but they are all artificially produced in particle accelerators. There is only one naturally occurring "radiocarbon").

Libby apparently first openly mentioned the ultimate purpose of his work on ^{14}C at a party (perhaps a Christmas party) in late 1946. This was when James R. Arnold, a recent Princeton Ph.D. and then postdoctoral student who, since February 1946, had been working with Libby (not on ^{14}C topics) and two other professors at Chicago, became cognizant of Libby's serious intention to use ^{14}C as a means of dating (Marlowe, 1980:1005; J. R. Arnold, personal communication). During the Christmas holidays, Arnold mentioned Libby's work to his father, who although an attorney by profession, was very knowledgeable in Egyptian archaeology. The elder Arnold, in turn communicated the news of this potential new dating technique to Ambrose Lansing, then the Director of the Metropolitan Museum in New York. The result was the first set of unsolicited ^{14}C samples to be received by a ^{14}C laboratory—sent to Libby at Chicago in January 1947 (Marlowe, 1980:1005–1006). That package contained what was to become the first sample to be dated by ^{14}C (C-1), cypress wood from the Sakkara tomb of the third dynasty Egyptian king Zoser (Djoser).

The arrival of these samples was certainly premature for as yet there was no experimental confirmation of any of Libby's assumptions concerning the distribution of natural ^{14}C in nature. With the collaboration of Grosse, who had worked with Libby on the Manhattan Project, and with the assistance of Libby's first graduate student at Chicago, Ernest C. Anderson, Libby moved in early 1947 to conduct the critical experiment to test the prediction he had made in his *Physical Review* paper. A sample of biological methane gas ("biomethane") was collected personally by Grosse from the sewage disposal plant in Baltimore, Maryland. Petroleum methane ("petromethane") was obtained from a Sun Oil Company refinery, whose president was a friend of Grosse. Both samples of methane were artificially enriched in Grosse's thermal diffusion column (used to make enriched ^{13}C for isotope tracer work in cancer research) at the Houdry Process Corporation laboratories at Marcus Hook, near Philadelphia, and the ^{14}C activity counted in a gas counter in Libby's lab (Libby, 1970a:2). The result was that whereas the ^{14}C activity of the biomethane increased in direct proportion to the measured amount of ^{13}C enrichment, there was no significant increase in the ^{14}C activity of the petromethane, indicating that the ^{14}C activity in the methane derived from the fossil source had long since decayed below detection limits. The success of this experiment initiated an intensive study by Anderson of the contemporary distribution of ^{14}C in nature and permitted Libby to make what apparently is the first specific published statement that ^{14}C might be used to determine the "ages of various carbonaceous materials" in the range of 1000–40,000 years (Anderson et al., 1947a:576; 1947b:936). The University of Chicago issued a news release on the day this article appeared. The text, as published in the New York Times (May 30, 1947), noted that "his [Libby's] discovery that radiocarbon has an atmospheric origin . . . will provide a new yardstick for measuring various earth periods. . . ."

6.3 LIBBY AMONG THE ARCHAEOLOGISTS

Up to that time, support for Libby's ^{14}C studies had been borne largely by internal funds from the University of Chicago. To expand the work, additional resources were needed. Through an improbable series of events, which included a lunchtime conversation between Libby's boss at the Manhattan Project and now a colleague at Chicago, Harold Urey, with Dutch paleoanthropologist G. H. R. von Koenigswald, Libby was put in touch with Paul Fejos, M.D. [Royal Hungarian Medical University], the Director of Research for the Viking Fund for Anthropological Research (now the Wenner-Gren Foundation). In November 1947, the Viking Fund initiated its support for ^{14}C dating research with a check made out to both Libby and Urey (Libby, 1980:1019; Marlowe, 1980:1007; cf. Dodds, 1973:100). With these funds Arnold, who, following his initial postdoctoral work at Chicago, had gone to Harvard on a National Research Council Fellowship, was asked to return to Chicago to be, in Libby's words, the "senior research man on the project—a physical chemist—[who] has a real interest in Egyptian archaeology" (Libby to Fejos, November 11, 1947, quoted in Marlowe, 1980:1008).

With Arnold's return to Chicago in late 1947, research on the ^{14}C method began to move rapidly forward. Although both Arnold and Anderson were involved in all aspects of the research, Anderson tended to concentrate on the development of the counting instrumentation and the question of the constancy of contemporary ^{14}C on a worldwide basis (his dissertation research), while Arnold focused on sample preparation problems for carbon and the dating of the first group of "known-age" samples. The need for known-age samples and the support provided by the Viking Fund provided the context for what apparently was the first formal presentation describing the method outside the confines of the University of Chicago. This was a lecture by Libby at a Viking Fund supper conference in New

York in January 1948 attended by about 30 individuals including anthropologists, archaeologists, and at least one geologist (G. Marlowe, personal
communication). In his invitation to Libby, Fejos cautioned that the "section of your talk dealing with physical chemistry should be on a popular
level as most of the anthropologists have little or no training in natural
sciences" (Fejos to Libby, November 17, 1947, quoted in Marlowe,
1980:1008).

A biographer of Paul Fejos reports that at the conclusion of Libby's
remarks, the audience remained mute until Richard Foster Flint, a Pleistocene geologist from Yale reportedly commented, "Well, if you people
are not interested in this, I am. . . . If you don't want anything dated, I
am for it, and would like to send some material" (Dodds, 1973:101).
Arnold, who was present at the meeting, while disputing the accuracy of
much of what was reported by Fejos' biographer as transpiring at the
meeting, does recall this remark by Flint. Arnold has suggested that one
possible explanation for the initial reaction of both the archaeologists
present as well as Flint's comments might have had something to do with
the relatively large sample sizes (i.e., pounds of carbon) required at this
point in the development of the technique. At this time, archaeologists,
and particularly those charged with the curation of museum collections,
typically would not have saved large amounts of charcoal or other organic
samples. They may have been extremely reluctant to turn over for de-
struction the valuable organic samples that they did have. Libby later
commented on this problem when he noted that "[t]hose museum dogs
were not going to give it to a bunch of physical chemists to burn up, no
way" (Libby, 1979a:43). Geologists, on the other hand, would more often
have access to larger samples of charcoal or wood and would not hesitate
to have them destroyed (J. R. Arnold, personal communication).

Another possible explanation for the reported initial silence of the archaeologists at this meeting could have been that they may not have completely understood Libby's presentation and were unclear as to the status
of the technique as it stood at that time. Frederick Johnson, who also
attended the meeting, reports that Libby had made a request for samples,
but that many who heard him misunderstood the nature of the request.
At this point in the development, *known-age* materials were needed to
test the basic ^{14}C model. In early January 1948, Libby was still not entirely
certain that there might not be some fatal defect in his reasoning. Johnson
reports that many archaeologists who heard Libby that night apparently
failed to appreciate this point (F. Johnson, personal communication;
Marlowe, 1980:1010).

Libby's initial interaction with archaeologists might also provide some
perspective on a puzzling statement apparently first publically made by
him during his Nobel Prize address in 1960. In his Nobel lecture he stated:

The research in the development of the dating technique consisted of two stages—the historical and the prehistorical epochs. The first shock Dr. Arnold and I had was when our advisors informed us that history extended back only to 5,000 years. We had thought initially that we would be able to get samples all along the curve back to 30,000 years, put the points in, and then our work would be finished. You read statements in books that such a society or archaeological site is 20,000 years old. We learned rather abruptly that these numbers, these ancient ages, are not known accurately; in fact, it is at about the time of the first dynasty in Egypt that the first historical date of any real certainty has been established. So we had, in the initial stages, the opportunity to check against knowns, principally Egyptian artifacts, and in the second stage we had to go into the great wilderness of prehistory to see whether there were elements of internal consistency which would lead one to believe that the method was sound or not [Libby, 1961a:102; see also Libby, 1961b:624].

Libby in fact had been informed of the approximately early third millenium B.C. historic boundary in the Near East as far back as 1946 (J. R. Arnold, personal communication). Perhaps up to the time of the Viking Fund dinner Libby might have assumed that archaeologists from areas other than the Near East could provide well-dated materials earlier than 3000 B.C. Another explanation is that Libby's 1960 rememberance may have reflected a penchant for dramatizing certain events in his retrospective oral presentations dealing with the development of ^{14}C dating.

6.4 THE CRITICAL PERIOD

The next 12-month period, essentially the whole of 1948, constituted a sort of micro-Manhattan project as far as the intensity of work on the ^{14}C method was concerned. What was required was a method to directly measure natural ^{14}C levels without resorting to costly enrichment procedures using a thermal diffusion column (Libby, 1970a:4). Although the use of proportional gas counters using methane or carbon dioxide was considered (Libby, 1947; E. C. Anderson, personal communication), this approach was immediately rejected in favor of the screen-wall counter design using elemental carbon. In early 1948, however, Libby was still not sure that the solid carbon method could be successfully developed to the point where it could be routinely employed to measure natural ^{14}C concentrations in samples. In the event that this could not be accomplished, he was prepared to use a thermal diffusion column for his ^{14}C work or move his counting equipment to a mine (Libby, 1972:xxxiv; Libby, 1979a:38–39). Arnold, with the assistance of a Houdry Process Corporation engineer, actually built a thermal diffusion column with a large portion of the Viking Fund grant and performed preliminary experiments. It was never routinely used for ^{14}C because of the subsequent success of the solid carbon approach (J. R. Arnold, personal communication).

A screen-wall counter design was chosen by the Chicago group for the initial series of experiments on the basis of several considerations. The principal problem was to fabricate a detector that could obtain the maximum and most stable count rates with the minimum physical size (Anderson, 1949, 1953; Kulp, 1954a). The physical size of the counters was crucial since the background rate of a counter is proportional to its size. It was also extremely important to employ a detector that yielded stable count rates over relatively long periods since attempts would be made to measure ^{14}C at natural concentrations close to the background levels of the counters. A detailed comparison of gas and screen wall detector designs determined that the counting sensitivities of the two alternative approaches was not significantly different for counter volumes and pressures that were then considered practical. (The use of larger counters and higher pressures later shifted the balance in favor of the gas counter.) The low efficiency of a screen-wall counter (about 5%) was offset by the fact that it could contain about 20 times more sample than a gas counter of similar size (Anderson et al., 1951; Anderson, 1953:74; E. C. Anderson, personal communication).

An important reason for the relative stability of the screen-wall counter was that it operated in the Geiger region. Because of the relatively high pulse heights, circuits could be made relatively insensitive to electronic noise. Gas counters operating in the proportional region required high-gain amplifiers with increased chances of instability (Anderson, 1949:40). Another investigator later determined that the screen-wall design increased the stability of the counter due perhaps to the fact that it produced a more uniform field effect throughout the detector's effective volume. Probably the most important factor, however, was that Libby obviously was very familiar with the operating characteristics of screen-wall counters through his association with them reaching back to 1931 (J. R. Arnold and E. C. Anderson, personal communications). It was not until severe problems surfaced beginning in the mid-1950s as a result of nuclear fallout contamination of sample preparations that solid carbon counting results were rendered questionable and the technique became obsolete (Anderson and Hayes, 1956:305; Anderson and Libby, 1957:405).

The major design features of the counter assembly used at the Chicago lab are shown in Fig. 6.2. Part (a) of this figure is essentially the diagram published in Libby's *Radiocarbon Dating;* part (b) identifies the important features of the design. The active volume of the detector (i.e., the portion within which the presence of ionizing radiation could be detected) represented about one-third of the total length of the cylinder as defined by

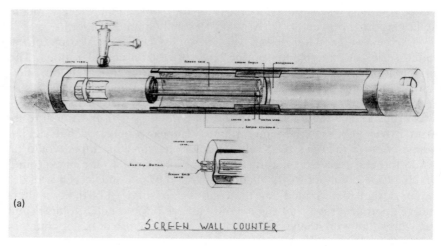

(a)

SCREEN WALL COUNTER

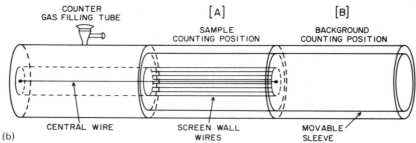

(b)

Figure 6.2 (a) Reproduction of a part of the original diagram of the Chicago screen wall counter (Libby, 1952a:54; 1955:57). (b) Simplified schematic of significant features of the counter. (Figure 6.2(a) as well as figures 6.3 through 6.7 have been made available to the author through the courtesy of Rainer Berger and Leona Marshall Libby, University of California, Los Angeles.)

the wire grid. An important outcome of this arrangement, as with the original Berkeley design, was to permit the background of the detector and ^{14}C activity of the sample to be measured repeatedly without changing the counting gas filling (Libby, 1967:12). This was accomplished in the same manner as in Libby's pre-World War II instrument—by having a sleeve that could be moved in and out of the sensitive volume of the counter. Samples intended for solid carbon counting were converted first to CO_2 by combustion or acidification (Fig. 6.3). Following chemical steps to eliminate trace impurities in the sample gas, the CO_2 was reduced by reaction with magnesium at high temperature. One of the reaction products was elemental carbon or "lamp black." The magnesium oxide was

Figure 6.3 Combustion line in the laboratory of W. F. Libby, Institute of Nuclear Studies, University of Chicago.

separated from the carbon by treating the reaction product with concentrated hydrochloric acid. Unfortunately, not all of the magnesium oxide—the "ash"—could be removed even with repeated acid treatments. The amount of ash remaining in the sample had to be determined carefully since it would reduce the observed ^{14}C specific activity (Libby, 1955:50–53).

The inside surface of the sample sleeve, constructed of metal rather than glass as was the case in the Berkeley instrument, was coated with the carbon powder, to which had been added, at various times, water, ethyl alcohol, and/or agar to form a paste. Figure 6.4 is a photograph of one of the sleeves with its coating of elemental carbon. The air used for

Figure 6.4 Sleeve of screen-wall counter with an elemental carbon coating. At one time, the Chicago laboratory used four counter sleeves which were given the names of "Matthew, Mark, Luke, and John"—so named because they gave the "gospel truth" (E. C. Anderson, personal communication). Later, two other sleeves named "Peter" and "Paul" were used (J. R. Arnold, personal communication).

evaporating off the water or alcohol contained radon, which resulted in unwanted radioactivity being picked up in the carbon sample (Libby, 1979a, p. 44). (Fortunately, radon has a relatively short half-life). After drying, the carbon coated sleeve was then fitted into the counter assembly. The cylinder was sealed with a special wax compound and evacuated. The counting gas was then introduced. Alternate sample and background counts were obtained by sliding the sleeve containing the sample coating in and out of the sensitive volume of the counter as defined by the wire grid.

A crucial aspect in the development of routine low-level ^{14}C work was the application of the anticoincidence principle to reduce background count rates in the detectors. This involved surrounding the central detector containing the sample with a ring of Geiger counters. With such an arrangement, pulses coming from the sample counter could be compared as to coincidence or noncoincidence (anticoincidence) with pulses coming from one or more of the "guard" counters. An electronic circuit could then be arranged so as to count pulses from the central counter only if these pulses were *not* accompanied by an essentially simultaneous pulse from the surrounding guard counters (see Fig. 4.7 in Chapter 4). Figure 6.5 represents a sample counter surrounded by the ring of GM counters used in the Chicago laboratory. Although Libby was not sure of the source of this approach (cf. Libby, 1970a:5; Libby, 1979a:39), he noted the principle of *coincidence* counting had been a standard method employed in cosmic-ray studies since the 1930s (cf. Anderson, 1953:75). It was being used in cyclotron experiments at Chicago at this time. This may have been the origin of the idea (L. M. Libby, n.d.; E. C. Anderson, personal communication).

Throughout 1948, problems with the counting instrumentation were slowly and painstakingly solved by Anderson and Arnold to the point that reasonably reliable results increasingly could be obtained (Anderson, 1949:ii). The effect of the various strategies used to shield the main detector from external radiation effects can be appreciated if one notes the various counting rates encountered during the Chicago experiments. Libby reported the background rate for an unshielded screen-wall counter at about 500 counts per minute (cpm). Placing the detector in an 8-in. steel shield assembly reduced the rate to about 100 cpm (Fig. 6.6). The use of the anticoincidence principle and numerous improvements in counter materials and experimental arrangements over a period of 15 months finally reduced the background rate to about 5–6 cpm with modern carbon counting about 12 cpm (Libby, 1967:10–11). The unremitting effort and close attention to detail needed to make the solid carbon technique yield consistent, accurate values is a tribute to the ability of the pioneering researchers.

Figure 6.5 Detail of screen-wall counter surrounded by ring of GM tubes. Compare with Libby (1955:68). The sample in this counter at the time when this photograph was taken was C-337, a sample of peat from a pollen zone identified as the "younger Allerod" from a site in northwestern Germany. The age listed in Libby (1955:86) was 11,044 ± 500.

By the end of 1948, the remaining nagging worry that there was a major flaw in Libby's model of how radiocarbon would function as a dating tool was essentially laid to rest (Libby *et al.*, 1949). Anderson's and Arnold's work with the recalcitrant counters permitted estimates of the specific ^{14}C activity of carbon to about ±10%. This was sufficient to determine

Figure 6.6 The shield used at the University of Chicago for solid carbon counting. This is a modification of the original design illustrated in Libby, 1952a, figure 10. The guard ring and sample counter can be seen in the interior cavity of the shield.

that one of the critical assumptions of the original concept was indeed essentially correct, namely that ^{14}C was reasonably well-mixed on a world-wide scale and therefore that modern biospheric ^{14}C activity should be constant. These data constituted the core of Anderson's dissertation, filed in late May 1949 (Anderson, 1949). The average terrestrial biospheric specific activity of ^{14}C reported by him was 12.50 ± 0.20 disintegrations per minute (dpm) per gram of carbon. This same average value was also reported in Libby *et al.* (1949), although there is some variation in the values cited for individual samples. A higher value of 15.3 ± 0.1 dpm per gram of carbon reported in Anderson and Libby (1951:69) and incorporated in Libby's *Radiocarbon Dating* (Libby, 1952a, 1955) was a result of the recalibration of the counters used by Anderson to account for the effect of differences in ash content in samples and counter efficiency (Libby, 1952a:16, footnote 4; E. N. Anderson, personal communication). Interestingly enough, however, even this first set of results revealed variations in ^{14}C activity of marine shell and in samples from the polar regions. Both of these problems continue to be studied down to the present day. Of historic interest also is that the first two radiocarbon "dates" were expressed not in years before present but in specific ^{14}C activity.

The first reported ^{14}C determination (C-1) was on the piece of cypress wood from the tomb of Djoser (Zoser) sent prematurely to Libby by Ambrose Lansing in January 1947. Its measured specific ^{14}C activity was less than one standard deviation from the expected value. The second sample to be measured was originally planned to have a known age about one-half that of the Zoser wood. However, the results of this measurement were never reported in print until 1967 (Libby, 1967:17). This is because its count rate was essentially the same as that of the contemporary samples measured by Anderson. What had happened was that John Wilson, a highly respected University of Chicago Egyptologist, had been asked to supply a piece of wood alledgedly from the Hellenistic period. He apparently did not know that the wood selected had originally been obtained from an antiquities dealer in Egypt. It in no sense was a "known age" sample. Its ^{14}C activity indicated that it was modern wood, i.e., a fake (Arnold in Marlowe, 1980:1012–1013). Apparently Professor Wilson, like some of the archaeologists at the Viking Fund Dinner, did not appreciate the fact that these were critical samples being used to *test* the ^{14}C method. Since Arnold reports that this affair ruined his Christmas (December, 1948), one wonders what would have happened if the wood supplied by John Wilson had been the first "known age" sample to be measured or if the sample supplied by Lansing had been a modern fake. Libby later commented that fortunately their early studies encountered few fakes "as otherwise faith in radiocarbon dating would have been rapidly shaken and the research abandoned. . . .'' He later publically mused that much of his ^{14}C research benefited from what he called "good fortune," which he suggested bordered on the "miraculous" (Libby, 1967:9, 17; Deevey, 1984:footnote 1).

6.5 KNOWNS AND UNKNOWNS

By the time of the Viking Fund dinner, his experiences with several potential submitters of samples may have convinced Libby of the need to provide his group with assistance to help obtain "known age" samples. In February 1948, as a direct outgrowth of discussions held during the Viking Fund meeting, the Executive Board of the then newly reorganized American Anthropological Association officially appointed a "Committee on Radioactive Carbon 14" (or "Committee on Carbon 14") which originally consisted of Frederick Johnson (R. S. Peabody Foundation) as chairman, Donald Collier (Director of the Field Museum of Natural History in Chicago), and Froelich Rainey (Director of the University of Pennsylvania Museum). The next month the Geological Society of America appointed Richard Foster Flint of Yale University to represent geological interests and the committee became a joint undertaking of both organizations (Johnson, 1951:2; Libby, 1952:v).

The capstone to the more than fifteen months of intensive work was "Age determinations by radiocarbon content: checks with samples of known age," published in the December 23, 1949 issue of *Science* (Arnold and Libby, 1949). In this paper, the first "curve of knowns" was presented—six data points (using seven samples) spanning a period from about A.D. 600 to 2700 B.C. Earlier in that year, Libby had led a group at the Argonne National Laboratory near Chicago to obtain a remeasurement of the ^{14}C half-life (Engelkemeir *et al.*, 1949). Their value of 5720 ± 47 years was used to convert the sample count rate of each sample to its radiocarbon age. The close (±10%) correspondence between the "known-age" and "radiocarbon age" of these seven samples indicated that the major assumption of the technique was valid for at least the last 5000 years to a typical precision of the measurements then possible, which was also about ±10%. The success with the first suite of known-age samples was the occasion for presentations to larger audiences of anthropologists and archaeologists, by Arnold in a paper delivered at a meeting of the Society for American Archaeology in May 1949 at the University of Indiana (*American Antiquity* 15:171) and by Libby in September at the XXIX International Congress of Americanists held at the American Museum of

Natural History in New York (see Bushnell, 1961). One wonders if the title of the *New York Times* (September 6, 1949) article "Scientist stumbles upon method to fix age of earth's materials" accurately reflected the manner in which Libby reported on his studies.

By early 1949, Libby had been confident enough to move into the "great unknown periods of prehistory." Since there were no fixed temporal points with which the ^{14}C values could be directly compared, it was hoped that validation would be accomplished by the degree of "internal consistency from a wide variety of samples and in a wide variety of problems (Arnold and Libby, 1949:680). The same strategy used to secure the known-age materials was employed, namely, assembling a group of collaborators. The membership of this group, selected officially by the AAA/GSA committee, was announced in the pages of *American Antiquity* in July 1949 (Griffin, 1949). Ten collaborators were named initially; two were added within a few months. Each collaborator was responsible for obtaining samples relevant to a particular region or chronological issue (Arnold and Libby, 1949:680; Johnson, 1951). Libby later described himself as an "amateur archaeologist" (Libby in Olsson, 1970a:107) presumably as a result of his involvement with his collaborators and the nature of most of the samples measured at Chicago.

During the period of the Chicago laboratory's routine dating operation, i.e., essentially from about March 1949 until early 1954 (when Libby was appointed a Commissioner of the U.S. Atomic Energy Commission by President Eisenhower), ^{14}C values on more than 500 samples were obtained. About two-thirds of these determinations were on samples of archaeological significance, most of which had been submitted by or through the collaborators. The first listing of "provisional" ^{14}C dates was prepared in mimeograph form on January 1, 1950, with a supplement dated in April. Both lists were sent principally to the collaborators. A number of these values were subsequently modified and these revised values appeared in a booklet published by the Institute of Nuclear Studies dated September 1, 1950. This booklet contained 148 ^{14}C values and carried a notation in the preface that the "list itself is not for publication in its present form, though the dates themselves may be quoted freely" (Arnold and Libby, 1950:1; Johnson 1951:3).

The text of the booklet with some modifications constituted the first Chicago list of dates, which appeared in *Science* in February 1951 (Arnold and Libby, 1951). This inaugurated the custom of laboratories publishing their ^{14}C values in the form of "date lists." For United States laboratories, the earliest date lists appeared in *Science*. The first Chicago date list included seven determinations that had not appeared in the 1950 booklet and excluded two. One of the excluded dates (C-276) appeared in the second Chicago list (Libby, 1951). However, the second excluded date never did reappear in any of the three subsequent Chicago date lists (Libby, 1952b, 1954a, 1954b), in any edition of *Radiocarbon Dating* (Libby, 1952a, 1955, 1965a), or in any subsequent index of dates (e.g., Deevey *et al.*, 1967). The only still unpublished Chicago ^{14}C value was a charcoal sample which was expected to be of Mousterian age. The average ^{14}C age cited in Arnold and Libby (1950:4) was 973 ± 230 ^{14}C years B.P. The comment in the text was that the date was "much too young." An examination of the correspondence files of the Chicago laboratory by the author has determined that the date was withdrawn at the last minute from the galleys of the first Chicago date list at the request of the archaeologist transmitting the sample. He feared that his relationship with the archaeologists in France who collected the sample would be jeopardized (cf. McBurney, 1952:40).

As we have noted, the half-life chosen by Libby to calculate the ^{14}C age determinations beginning with the Chicago date list booklet in 1950 was 5568 ± 30 years. This value replaced the earlier value of 5720 ± 47, which had been used to calculate the first "curve of knowns" (Arnold and Libby, 1949). The 5568 ± 30 years value became known as the "Libby half-life" and has continued to be used to calculate "conventional" radiocarbon age values even though it now is clear that the half-life Libby originally used (5720 ± 47) was probably more nearly correct. With only a few exceptions, the 5568 ± 30 half-life has been used to calculate all of the ^{14}C values published in the journal *Radiocarbon*.

6.6 RADIOCARBON DATING COMES OF AGE

The acceptance by the vast majority of scientists from a number of disciplines of the general validity of the ^{14}C method is reflected in the rapid establishment of laboratories to perform ^{14}C analyses. Solid carbon counting instrumentation was installed at eight institutions in the early 1950s. These pioneering laboratories included the University of Arizona (Wise and Shutler, 1958), University of Copenhagan, Denmark (Anderson *et al.*, 1953), Lamont Geological Observatory, Columbia University (Kulp *et al.*, 1951), University of Michigan (Crane, 1956), New Zealand Institute of Nuclear Studies (Fergusson and Rafter, 1953), University of Pennsylvania (Ralph, 1955), University of Saskatchewan (McCallum, 1955), and Yale University (Blau *et al.*, 1953). In at least three instances at United States universities—Arizona, Michigan, and Pennsylvania—archaeologists were instrumental in establishing the laboratories. Some thought at that moment that they stood "before the threat of the atom in the form of C14 dating. This may be our last chance for old-fashioned, uncontrolled guessing" (Phillips *et al.*, 1951:455).

Low-level ^{14}C counting using Libby's solid carbon approach required an extraordinary attention to detail in laboratory procedure. Many who attempted to duplicate the technique experienced moderate to severe difficulties in obtaining reproducible values. Most frustrating was the problem of avoiding contamination of the solid carbon samples with nuclear fission fallout products resulting from the testing of thermonuclear devices in the atmosphere. As a result of these and other problems, the laboratories previously noted, which had begun operations with solid carbon counting, converted to some type of gas counting by the end of the decade. Meanwhile, other early laboratories had commenced operations with gas counting using CO_2 or acetylene. These included the facilities at the University of Groningen, Netherlands (de Vries and Barendsen, 1954), the University of Heidelberg, Germany (Munnich, 1957) and the United States Geological Survey in Washington, D.C. (Suess, 1954b).

By the time of the inauguration of the journal *Radiocarbon* in 1959, more than twenty ^{14}C laboratories were actively pursuing a whole range of studies. *Radiocarbon* (initially the *Radiocarbon Supplement* to the *American Journal of Science*) was inaugurated to solve the problem of the increasing number of radiocarbon laboratory date lists that could no longer be accommodated in *Science* as well as to provide an archive of primary data on individual ^{14}C dates (Deevey, 1984:2). This later function was of particular concern at that time since, in some cases, ^{14}C values were being cited without laboratory number or without descriptions of any kind (Deevey and Flint, 1959:preface). Rapid development during this period took place particularly in Europe (Waterbolk, 1960) as well as in labs being built in England (at the British Museum and Cambridge), in Sweden (Stockholm and Uppsala), France (Gif and Saclay), Ireland (Dublin), Germany (Cologne and Hanover), Belgium (Louvain), Russia (Moscow), Italy (Pisa and Rome), Switzerland (Bern) and Norway (Trondheim). In the United States during this period, the first commercial ^{14}C laboratory was opened and several oil companies briefly supported ^{14}C facilities (Brannon *et al.*, 1957a,b).

Initial ^{14}C data, which yielded age estimates at dramatic variance with the views of individual archaeologists or geologists (e.g., Neustupny, 1970; Antevs, 1957; cf. Lee, 1981), generated discussions—a number transmitted only orally—which tended to question the validity of the ^{14}C method in general (Barker, 1958; F. Johnson, personal communication). With the rapidly mounting evidence of the general validity of the ^{14}C model in broad outline, discussions turned to questions of accuracy of ^{14}C values from specific archaeological or geological contexts or geochemical environments (e.g., Broecker and Kulp, 1956). Interestingly enough, since the late 1950s, objections to the general accuracy of the method have been based almost entirely on religious or theological grounds most often expressed within a European or American Protestant fundamentalist/creationist apologetic framework (e.g., Pearl, 1963; Slusher, 1981; Brown, 1983; DeYoung, 1978; cf. Brown, 1986).

Despite this opposition, over the years a number of ^{14}C determinations have been carried out on samples of concern to this interest group. For example, wood reportedly extracted from a glacier at the 4200 m (about 14,000 foot) level on one of the volcanic peaks of Mount Ararat in northeastern Turkey were offered as evidence of the existence at that location of the remains of a large structure. It was argued that this feature represented the remnants of an ocean-going vessel, specifically a boat known in Biblical literature as the "Ark of Noah." In all, six radiocarbon laboratories over more than a decade have independently carried out ^{14}C measurements on wood alledgedly collected from this feature. Five of the six ^{14}C determinations were statistically identical at the 1σ level. Based on this data, the age of these samples was placed in the sixth to ninth century A.D. It was suggested that the wood may represent the remnants of a cenotaph or memorial erected by Armenian or Byzantine clerics to commemorate the location of what they believed was the final resting place of the Ark of Noah. Perhaps the cenotaph was actually built in the form of a boat (Taylor and Berger, 1980).

The capstone at the end of the first decade of ^{14}C dating was the award of the 1960 Nobel Prize in Chemistry to Libby for the development of the method (Fig. 6.7). Translated from Swedish, the citation reads:

> At its meeting of November 3, 1960, the Royal Swedish Academy of Science has decided, in conformity with the terms of the November 27, 1895 will of Alfred Nobel, to award the prize to be given this year for the most important chemical discovery or improvement to Willard F. Libby for his method to use Carbon-14 for age determinations in archaeology, geology, geophysics, and other sciences.

(The author is grateful to Professor Gorman Bjorch, University of Stockholm and Professor Tord Ganelius, Secretary of the Royal Swedish Academy of Science for the translation of the Swedish text.) At the time the Nobel Prize was awarded, Libby had recently arrived to take up a position of Professor of Chemistry and Director of the Institute of Geophysics and Planetary Physics (IGPP) at the University of California, Los Angeles (UCLA). Libby brought to UCLA Gordon J. Fergusson, formerly of the

*K*ungliga Svenska Vetenskaps-
akademien har vid sin samman-
komst den 3 november 1960 i enlig-
het med föreskrifterna i det av
ALFRED NOBEL
den 27 november 1895 upprättade
testamentet beslutat att överlämna
det pris som detta år bortgives för
den viktigaste kemiska upptäckt
eller förbättring till
WILLARD F. LIBBY
för hans upptäckt av kol-14 som
tidmätare inom arkeologi, geologi,
geofysik m.fl. vetenskaper.
Stockholm den 10 december 1960/

Akademiens preses   Akademiens sekreterare

Figure 6.7   The text of the 1960 Nobel Prize in Chemistry awarded to Willard F. Libby for the development of the ^{14}C dating technique. See text for English translation from original Swedish. (Original made available by Leona Marshall Libby.)

New Zealand Institute of Nuclear Studies ^{14}C laboratory, to build a radiocarbon dating laboratory for the IGPP and collaborate in geophysical research. Libby retired from UCLA in 1976 and died in Los Angeles at the age of 71 on September 11, 1980 (Libby, L. M., 1981; Burleigh, 1981; Berger, 1983).

REFERENCES

Aitken, M. J.
  1974   Radiocarbon dating. In *Physics and archaeology* (2nd ed.). Oxford: Clarendon
          Press. Chap. 2.
  1985   *Thermoluminescence dating*. London: Academic Press.
Albero, M. C., and F. E. Angiolini
  1985   INGEIS radiocarbon laboratory dates II. *Radiocarbon 27*:314–337.
Anbar, M.
  1978   The limitations of mass spectrometric radiocarbon dating using CN⁻ ions. In
          Proceedings of the First Conference on Radiocarbon Dating with Accelerators,
          edited by H. E. Gove. Rochester, New York: University of Rochester. Pp.
          152–155.
Anderson, E. C.
  1949   Natural radiocarbon. Unpublished Ph.D. dissertation, University of Chicago.
  1953   The production and distribution of natural radiocarbon. *Annual Review of Nu-
          clear Science 2*:63–89.
Anderson, E. C., J. R. Arnold, and W. F. Libby
  1951   Measurement of low level radiocarbon. *Review of Scientific Instruments 22*:225–
          230.
Anderson, E. C., and R. N. Hayes
  1956   Recent advances in low level counting techniques. *Annual Review of Nuclear
          Science 6*:303–323.
Anderson, E. C., H. Levi, and H. Tauber
  1953   Copenhagen natural radiocarbon measurements, I. *Science 118*:6–9.
Anderson, E. C., and Libby, W. F.
  1951   World-wide distribution of natural radiocarbon. *Physical Review 81*:64–69.
  1957   The development and applications of low level counting. *Advances in Biological
          and Medical Physics 5*:385–411.

Anderson, E. C., W. F. Libby, S. Weinhouse, A. F. Reid, A. D. Kirshenbaum, and A. V.
        Grosse
    1947a   Radiocarbon from cosmic radiation. *Science 105:*576.
    1947b   Natural radiocarbon from cosmic radiation. *Physical Review 72:*931–936.
Andrews, E. W., V
    1978   Endnote: The northern Maya lowlands sequence. In *Chronologies in New World
            archaeology,* edited by R. E. Taylor and C. W. Meighan. New York: Academic
            Press. Pp. 377–381.
Antevs, E.
    1957   Geological tests on the varve and radiocarbon chronologies. *Journal of Geology
            65:*129–148.
Arnold, J. R.
    1954   Scintillation counting of natural radiocarbon: I. The Counting Method. *Science
            119:*155–157.
Arnold, J. R., and W. F. Libby
    1949   Age determinations by radiocarbon content: checks with samples of known
            age. *Science 110:*678–680.
    1950   *Radiocarbon dates (September 1, 1950).* Chicago: University of Chicago, In-
            stitute for Nuclear Studies.
    1951   Radiocarbon dates. *Science 113:*111–120.
Arundale, W. H.
    1981   Radiocarbon dating in eastern Arctic archeology: A flexible approach. *American
            Antiquity 46:*244–271.
Ashmore, P. J., and P. H. Hill
    1983   Broxmouth and high-precision dating. In *Archaeology, dendrochronology and
            the radiocarbon calibration curve,* edited by R. S. Ottaway. Occasional Paper
            No. 9, Department of Archaeology, University of Edinburgh. Pp. 83–98.
Atkinson, R. J. C.
    1975   British prehistory and the radiocarbon revolution. *Antiquity 49:*173–177.
Audric, B. N., and J. V. P. Long
    1954   Use of dissolved acetylene in liquid scintillation counters for the measurement
            of carbon-14 for low specific activity. *Nature (London) 173:*992–993.
Bailey, D. K.
    1970   Phytogeography and taxonomy of *Pinus* subsection *Balfourianae. Annals of
            the Missouri Botanical Garden 57:*210–249.
Bailey, J. M., and R. Lee
    1973   The effect of alkaline pretreatment on the radiocarbon dates of several New
            Zealand charcoals. In *Proceedings of the Eighth International Radiocarbon
            Dating Conference,* compiled by T. A. Rafter and T. Grant-Taylor. Wellington:
            Royal Society of New Zealand. Pp. 582–591.
Baillie, M. G. L., J. R. Pilcher, and G. W. Pearson
    1983   Dendrochronology at Belfast as a background to high-precision calibration.
            *Radiocarbon 25:*171–178.
Bannister, B., and P. E. Damon
    1973   A dendrochronologically-derived primary standard for radiocarbon dating. In
            Proceedings of the Eighth International Radiocarbon Dating Conference,
            compiled by T. A. Rafter and T. Grant-Taylor. Wellington: Royal Society of
            New Zealand. Pp. 676–685.

Barbetti, M. F.
1980   Geomagnetic strength over the last 50,000 years and changes in atmospheric ^{14}C concentration: Emerging trends. *Radiocarbon 22:*191–199.

Barbetti, M. F., and M. W. McElhinney
1972   The Lake Mungo geomagnetic excursion. *Philosophical Transactions of the Royal Society of London A281:*515–542.

Barendsen, G. W.
1957   Radiocarbon dating with liquid CO_2 as diluent in a scintillation solution. *Review of Scientific Instruments 28:*430–432.

Barker, H.
1953   Radiocarbon dating: Large-scale preparation of acetylene from organic material. *Nature (London) 172:*631–632.
1958   Radio carbon dating: Its scope and limitations. *Antiquity 32:*253–263.
1970   Critical assessment of radiocarbon dating. *Philosophical Transactions of the Royal Society of London A269:*23–26.

Barnard, N.
1982   Radiocarbon dating and other laboratory derived data: The historian's view-point. In *Archaeometry: An Australasian perspective,* edited by W. Ambrose and P. Duerden. Canberra: Department of Prehistory, Research School of Pacific Studies, Australian National University. Pp. 357–360.

Barton, C. E., R. T. Merrill, and M. Barbetti
1979   Intensity of the earth's magnetic field over the last 10,000 years. *Physics of the Earth and Planetary Interiors 20:*96–111.

Baxter, M. S.
1983   An international tree ring replicate study. In *^{14}C and archaeology,* edited by W. G. Mook and H. T. Waterbolk. Strasbourg: Council of Europe. Pp. 123–132.

Baxter, M. S., and J. G. Farmer
1973a   Glasgow University radiocarbon measurements VII. *Radiocarbon 15:*488–492.
1973b   Radiocarbon: short-term variations. *Earth and Planetary Science Letters 20:*295–299.

Baxter, M. S., and A. Walton
1970   Radiocarbon dating of mortars. *Nature (London) 225:*937–938.

Becker, B.
1983   The long-term radiocarbon trend of the absolute German oak tree-ring chronology, 2800 to 800 B. C. *Radiocarbon 25:*197–203.

Beer, J., M. Andree, H. Oeschger, U. Siegenthaler, G. Bonani, H. Hofmann, E. Morenzoni, M. Nessi, M. Suter, W. Wolfli, R. Finkel, and C. Langway, Jr.
1984   The Camp Century ^{10}Be record: Implications for long-term variations of the geomagnetic dipole moment. *Nuclear Instruments and Methods in Physics Research 233*(B5):380–384.

Beer, J., M. Andree, H. Oeschger, B. Stauffer, R. Balzer, G. Bonani, C. Stoller, M. Suter, W. Wolfli, and R. C. Finkel
1983   Temporal ^{10}Be variations in ice. *Radiocarbon 25:*269–278.

Beer, J., V. Giertz, M. Holl, H. Oeschger, T. Reisen, and C. Strahm
1979   The contribution of the Swiss lake-dwellings of the calibration of radiocarbon dates. In *Radiocarbon dating,* edited by R. Berger and H. E. Suess. Los Angeles: University of California Press. Pp. 566–590.

Bender, M. M.
   1968   Mass spectrometric studies of carbon 13 variations in corn and other grasses.
            *Radiocarbon 10:*468–472.
   1971   Variations in the $^{13}C/^{12}C$ ratios of plants in relation to the pathway of photo-
            synthetic carbon dioxide fixation. *Phytochemistry 10:*1239–1244.
Bender, M. M., I. Rouhani, H. M. Vines, and C. C. Black, Jr.
   1973 $^{13}C/^{12}C$ ratio changes in Crassulacean acid metabolism plants. *Plant Physiology
            52:*427–430.
Bennett, C. L.
   1979   Radiocarbon dating with accelerators. *American Scientist 67:*450–457.
Bennett, C. L., R. P. Beukens, M. R. Clover, D. Elmore, H. E. Gove, L. Kilius, A. E.
            Litherland, and K. H. Purser
   1978   Radiocarbon dating with electrostatic accelerators: Dating of milligram samples.
            *Science 201:*345–347.
Bennett, C. L., R. P. Beukens, M. R. Clover, H. E. Gove, R. B. Liebert, A. E. Litherland,
            K. K. Purser, and W. E. Sondheim
   1977   Radiocarbon dating using accelerators: Negative ions provide the key. *Science
            198:*508–509.
Berger, R.
   1970a   Ancient Egyptian radiocarbon chronology. *Philosophical Transactions of the
            Royal Society of London A269:*23–36.
   1970b   The potential and limitations of radiocarbon dating in the Middle Ages: The
            radiochronologist's view. In *Scientific methods in medieval archaeology,* edited
            by R. Berger. Berkeley: University of California Press. Pp. 89–139.
   1973   Tree-ring calibration of radiocarbon dates. In *Proceedings of the Eighth In-
            ternational Radiocarbon Dating Conference,* compiled by T. A. Rafter and
            T. Grant-Taylor. Wellington: Royal Society of New Zealand. Pp. A97–A103.
   1979   Radiocarbon dating with accelerators. *Journal of Archaeological Science 6:*101–
            104.
   1983   Willard Frank Libby 1908–1980. In *^{14}C and archaeology,* edited by W. G. Mook
            and H. T. Waterbolk. [*PACT 8:*13–16]. Strasbourg: Council of Europe. Pp. 13–16.
   1985   Suess' "wiggles and deviations" proven by historical and archaeological means.
            *Meteoritics 20:*395–401.
Berger, R., G. J. Fergusson, and W. F. Libby
   1965   UCLA radiocarbon dates IV. *Radiocarbon 7:*336–371.
Berger, R., A. G. Horney, and W. F. Libby
   1964   Radiocarbon dating of bone and shell from their organic components. *Science
            144:*999–1001.
Berger, R. and W. F. Libby
   1967   UCLA radiocarbon dates VI. *Radiocarbon 9:*477–504.
Berger, R., and L. M. Libby (editors)
   1981   *Radiocarbon and tritium,* Vol. 1: *The Publications of Willard Libby.* Santa
            Monica: Geo Science Analytical, Inc.
Berger, R., and H. E. Suess (editors)
   1979   *Radiocarbon dating.* Proceedings of the Ninth International Conference, Los
            Angeles and La Jolla, 1976. Berkeley: University of California Press.
Berger, R., R. E. Taylor, and W. F. Libby
   1966   Radiocarbon content of marine shells from the California and Mexican west
            coast. *Science 153:*864–866.

Berglund, B. E., S. Håkansson, and E. Lagerlund
1976   Radiocarbon dated Mammoth (*Mammuthus primigenius* Blumenbach) finds in
           south Sweden. *Boreas 5:*177–191.
Berland, T.
1962   *The scientific life.* New York: Coward-McCann. Pp. 13–49.
Betancourt, J. L., A. Long, D. J. Donahue, A. J. T. Jull, and T. H. Zabel
1984   Native or alien? The case for North American *Corispermum L.* (Chenopodi-
           aceae). *Science 311:*653–655.
Bischoff, J. L., R. Merriam, W. M. Childers, and R. Protsch
1976   Antiquity of man in America indicated by radiometric dates on the Yuha burial
           site. *Nature (London) 261:*128–129.
Blau, M., E. S. Deevey, Jr., and M. S. Gross
1953   Yale natural radiocarbon Measurements, I. Pyramid Valley, New Zealand and
           its problems. *Science 118:*1–6.
Blong, R. J., and R. Gillespie
1978   Fluvially transported charcoal gives erroneous ages for recent deposits. *Nature
           (London) 271:*739–741.
Brannon, H. R., Jr., A. C. Daughtry, D. Perry, L. H. Simons, W. W. Whitaker, and M.
           Williams
1957b   Humble Oil Company radiocarbon dates I. *Science 125:*147–150.
Brannon, H. R., Jr., L. H. Simons, D. Perry, A. C. Daughtry, and E. McFarlan, Jr.
1957a   Humble Oil Company radiocarbon dates II. *Science 126:*191–923.
Broecker, W. S., R. G. M. Ewing, and B. C. Heezen
1960   Natural radiocarbon in the Atlantic Ocean. *Journal of Geophysical Research
           65:*2903–2931.
Broecker, W. S., and J. L. Kulp
1956   The radiocarbon method of age determination. *American Antiquity 22:*1–
           11.
Broecker, W. S., J. L. Kulp, and C. S. Tucek
1956   Lamont natural radiocarbon measurements III. *Science 129:*154–165.
Broecker, W. S., and E. A. Olson
1959   Lamont radiocarbon measurements VI. *American Journal of Science Radio-
           carbon Supplement 1:*111–113.
1961   Lamont radiocarbon measurements VIII. *Radiocarbon 3:*176–204.
Broeker, W. S., and T. Peng
1982   *Tracers in the Sea.* Palisades, New York: Lamont-Doherty Geological Ob-
           servatory, Columbia University.
Brothwell, D., and R. Burleigh
1977   On sinking Otavalo Man. *Journal of Archaeological Science 4:*291–294.
Browman, D. L.
1981   Isotopic discrimination and correction factors in radiocarbon dating. In *Ad-
           vances in archaeological method and theory,* edited by M. B. Schiffer. (Vol.
           4). New York: Academic Press. Pp. 241–295.
Brown, R. H.
1983   The interpretation of carbon-14 age data. In *Origin by design,* H. G. Coffin
           with R. H. Brown. Washington, D.C.: Review and Herald Publishing Asso-
           ciation. Pp. 309–329.
1986 ^{14}C depth profiles as indicators of trends in climate and $^{14}C/^{12}C$ ratio. *Radio-
           carbon 28:*350–357.

Bruns, M., I. Levin, K. O. Munnish, H. W. Hubberten, and S. Fillipakis
  1980   Regional sources of volcanic carbon dioxide and their influence on ^{14}C content
           of present-day plant material. *Radiocarbon 22:*532–536.
Bryan, A. L., R. M. Casamiquela, J. M. Cruxent, R. Gruhn, and C. Ochsenius
  1978   An El Jobo mastodon kill at Taima–taima, Venezuela. *Science 200:*1275–1277.
Bucha, V.
  1967   Archaeomagnetic and paleomagnetic study of the magnetic field of the earth
           in the past 600,000 years. *Nature (London) 213:*1005–1007.
  1970   Influence of the earth's magnetic field on radiocarbon dating. In *Radiocarbon
           variations and absolute chronology,* edited by I. U. Olsson. Stockholm:
           Almqvist & Wiksell. Pp. 501–511.
Burleigh, R.
  1974   A bomb method for rapid combustion of samples. In *Liquid scintillation count-
           ing,* edited by M. A. Crook and P. Johnson. (Vol. 3). London: Heyden. Pp.
           295–302.
  1981   W. F. Libby and the development of radiocarbon dating. *Antiquity 55:*96–98.
  1983   Two radiocarbon dates for freshwater shells from Hierakonpolis: Archaeological
           and geological interpretations. *Journal of Archaeological Science 10:*361–367.
Burleigh, R., and A. D. Baynes-Cope
  1983   Possibilities in the dating of writing materials and textiles. *Radiocarbon 25:*669–
           674.
Burleigh, R., and A. Hewson
  1979   Archaeological evidence for short-term natural ^{14}C variations. In *Radiocarbon
           dating,* edited by R. Berger and H. E. Suess. Berkeley: University of California
           Press. Pp. 591–600.
Burleigh, R., and M. P. Kerney
  1982   Some chronological implications of a fossil molluscan assemblage from a neo-
           lithic site at Brook, Kent, England. *Journal of Archaeological Science 9:*29–
           38.
Burleigh, R., K. Matthews, and M. M. Leese
  1984   Consensus ^{13}C values. *Radiocarbon 26:*46–53.
Bushnell, G.
  1961   Radiocarbon dates and new world chronology. *Antiquity 35:*286–291.
Butzer, K. W.
  1978   Toward an integrated, contextual approach in archaeology: A personal view.
           *Journal of Archaeological Science 5:*191–194.
Cahen, D., and J. Moeyersons
  1977   Subsurface movements of stone artifacts and their implications for the prehistory
           of central Africa. *Nature (London) 266:*812–815.
Cahen, D., J. Moeyersons, and W. G. Mook
  1983   Radiocarbon dates from Gombe Point (Kinshasa, Zaire) and their implications.
           In *^{14}C and archaeology,* edited by W. G. Mook and H. T. Waterbolk. [*PACT
           8:*441–453]. Strasbourg: Council of Europe. Pp. 441–453.
Cain, W. F., and H. E. Suess
  1976   Carbon 14 in tree rings. *Journal of Geophysical Research 81:*3688–3694.
Calf, G. E., and H. A. Polach
  1973   Teflon vials for liquid scintillation counting of carbon-14 samples. In *Liquid
           scintillation counting: Recent developments,* edited by P. E. Stanley and B.
           A. Scoggins. New York: Academic Press. Pp. 223–234.
Campbell, J. A., M. S. Baxter, and L. Alcock
  1979   Radiocarbon dates for the Cadbury massacre. *Antiquity 53:*31–38.

Campbell, J. M.
1965   Radiocarbon dating in far northern archaeology. In *Proceedings of the Sixth International Conference Radiocarbon and Tritium Dating*, compiled by R. M. Chatters and E. A. Olson. Springfield, Virginia: Clearinghouse for Federal Scientific and Technical Information. Pp. 179–186.

Carmi, I., M. Stiller, and A. Kaufman
1985   The effect of atmospheric ^{14}C variations on the ^{14}C levels in the Jordan River system. *Radiocarbon 27*:305–313.

Castagnoli, G., and D. Lal
1980   Solar modulation effects in terrestrial production of carbon-14. *Radiocarbon 22*:133–158.

Cavallo, L. M., and W. B. Mann
1980   New National Bureau of Standards contemporary carbon-14 standards. *Radiocarbon 22*:962–963.

Chappell, J., and H. A. Polach
1972   Some effects of partial recrystallization on ^{14}C dating of late Pleistocene corals and molluscs. *Quaternary Research 2*:244–252.

Chatters, R. M., J. W. Crosby, III, and L. G. Engstrand
1969   *Fumarole gaseous emanations: Their influence on carbon-14 dates*. Circular 32, College of Engineering, Washington State University.

Chatters, R. M., and E. A. Olson (compilers)
1965   *Proceedings of the Sixth International Conference Radiocarbon and Tritium Dating* [Conf-650652]. Springfield, Virginia: Clearinghouse for Federal Scientific and Technical Information.

Clark, G.
1961   *World prehistory*. Cambridge: Cambridge University Press.
1969   *World prehistory: A new outline* (2nd ed.). Cambridge: Cambridge University Press.
1970   *Aspects of prehistory*. Berkeley: University of California Press.

Clark, J. D.
1979   Radiocarbon dating and African archaeology. In *Radiocarbon dating*, edited by R. Berger and H. E. Suess. Berkeley: University of California Press. Pp. 7–31.
1984   Foreword. In R. Gillespie, *Radiocarbon user's handbook*. Oxford: Oxonian Rewley Press.

Clark, R. M.
1975   A calibration curve for radiocarbon dates. *Antiquity 49*:251–266.
1978   Bristlecone pine and ancient Egypt: a re-appraisal. *Archaeometry 20*:5–17.

Clark, R. M., and C. Renfrew
1972   A statistical approach to the calibration of floating tree-ring chronologies using radiocarbon dates. *Archaeometry 14*:5–19.

Clark, R. M., and A. Sowray
1973   Further statistical methods for the calibration of floating tree-ring chronologies. *Archaeometry 15*:255–256.

Conrad, N., D. L. Asch, N. B. Asch, D. Elmore, H. Gove, M. Rubin, J. A. Brown, M. D. Wiant, K. B. Farnsworth, and T. G. Cook
1984   Accelerator radiocarbon dating of evidence for prehistoric horticulture in Illinois. *Nature (London) 308*:443–446.

Cook, S. F.
1964   The nature of charcoal excavated at archaeological sites. *American Antiquity 29*:514.

Craig, H.
> 1953   The geochemistry of the stable carbon isotopes. *Geochimica et Cosmochimica Acta 3:*53–92.
> 1954   Carbon 13 in plants and the relationships between carbon 13 and carbon 14 variations in nature. *Journal of Geology 62:*115–149.
> 1957   Isotopic standards for carbon and oxygen and correction factors for mass spectrometric analysis of carbon dioxide. *Geochimica et Cosmochimica Acta 12:*133–149.
> 1961   Mass-spectrometer analyses of radiocarbon standards. *Radiocarbon 3:*1–3.

Crane, H. R.
> 1956   University of Michigan radiocarbon dates I. *Science 124:*664–672.

Currie, L. A., and H. A. Polach
> 1980   Exploratory analysis of the international radiocarbon cross-calibration data: Consensus values and interlaboratory error. *Radiocarbon 22:*933–935.

Damon, P. E., C. Haynes, and A. Long
> 1964   Arizona radiocarbon dates V. *Radiocarbon 6:*91–107.

Damon, P. E., J. C. Lerman, and A. Long
> 1978   Temporal fluctuations of atmospheric ^{14}C: Causal factors and implications. *Annual Review of Earth and Planetary Science 6:*457–494.

Damon, P. E. and T. W. Linick
> 1986   Geomagnetic–geliomagnetic modulation of atmospheric radiocarbon production. *Radiocarbon 28:*266–278.

Damon, P. E. and A. Long
> 1962   Arizona radiocarbon dates III. *Radiocarbon 4:*239–249.

Damon, P. E., A. Long, and E. I. Wallick
> 1973   Comments on "Radiocarbon: Short-term variations" by M. S. Baxter and J. G. Farmer. *Earth and Planetary Science Letters 20:*311–314.

Damon, P. E., R. Sternberg, and C. J. Radnell
> 1983   Modeling of atmospheric radiocarbon fluctuations for the past three centuries. *Radiocarbon 25:*249–258.

Daniel, G.
> 1959   Editorial. *Antiquity 33:*79–80.
> 1967   *The origins and growth of archaeology.* New York: Crowell.
> 1972   Editorial. *Antiquity 46:*265.

Dauchot-Dehon, M., M. Van Strydonck, and J. Heylen
> 1983   Institut Royal du Patrimoine Artistique radiocarbon dates IX. *Radiocarbon 25:*867–874.

Davis, E. M.
> 1965   Interdisciplinary appraisal of radiocarbon dates in archeology. In *Proceedings of the Sixth International Conference Radiocarbon and Tritium Dating,* compiled by R. M. Chatters and E. A. Olson. Springfield, Virginia: Clearinghouse for Federal Scientific and Technical Information. Pp. 294–303.

Dean, J. S.
> 1978   Independent dating in archaeological analysis. In *Advances in archaeological method and theory,* edited by M. B. Schiffer. New York: Academic Press. Vol. 1, pp. 223–265.
> 1985   Review of *Time, space, and transition in Anasazi prehistory* by M. S. Berry. *American Antiquity 50:*704.

De Atley, S. P.
> 1980   Radiocarbon dating of ceramic materials: Progress and prospects. *Radiocarbon 22:*987–993.

Deevey, E. S., Jr.
1984   Zero BP plus 34: 34 years of radiocarbon. *Radiocarbon 26:1–6.*
Deevey, E. S., Jr., and R. F. Flint
1959   Preface. *Radiocarbon 1:1.*
Deevey, E. S., Jr., R. F. Flint, and I. Rouse
1967   *Radiocarbon measurements: Comprehensive index, 1950–1965.* New Haven: Yale University.
Deevey, E. S., Jr., L. J. Granlenshi, and V. Hoffren
1959   Yale natural radiocarbon measurements IV. *Radiocarbon 1:144–172.*
Deevey, E. S., Jr., M. A. Gross, G. W. Hutchinson, and H. L. Kraybill
1954   The natural C^{14} contents of materials from hard-water lakes. *Proceedings of the National Academy of Sciences (U.S.A.) 40:285–288.*
Delibrias, G., M. T. Guillier, and J. Labeyrie
1974   Gif natural radiocarbon measurements VIII. *Radiocarbon 16:15–94.*
Delibrias, G., and J. Labeyrie
1965   The dating of mortars by the carbon-14 method. In *Proceedings of the Sixth International Conference Radiocarbon and Tritium Dating,* compiled by R. M. Chatters and E. A. Olson. Springfield, Virginia: Clearinghouse for Federal Scientific and Technical Information. Pp. 344–346.
Dodds, John W.
1973   *The several lives of Paul Fejos.* New York: Wener-Gren Foundation.
Donahue, D. J., A. J. T. Jull, and T. H. Zabel
1984   Results of radioisotope measurements at the NSF–University of Arizona tandem accelerator mass spectrometer facility. *Nuclear Instruments and Methods in Physics Research 233*(B5):162–166.
Donahue, D. J., A. J. T. Jull, T. H. Zabel, and P. E. Damon
1983   The use of accelerators for archaeological dating. *Nuclear Instruments and Methods in Physics Research 218:435–429.*
Dyck, W. and J. G. Fyles
1962   Geological survey of Canada radiocarbon dates I. *Radiocarbon 4:13–26.*
1964   Geological survey of Canada radiocarbon dates III. *Radiocarbon 6:167–181.*
1965   Geological survey of Canada radiocarbon dates IV. *Radiocarbon 7:24–46.*
Dyck, W., J. A. Lowdon, J. G. Fyles, and W. Blake, Jr.
1966   Geological survey of Canada radiocarbon dates V. *Radiocarbon 8:96–127.*
Edwards, I. E. S.
1970   Absolute dating from Egyptian records and comparison with carbon-14 dating. *Philosophical Transactions of the Royal Society of London A269:11–18.*
Engelkemeir, A. G., W. H. Hamill, M. G. Inghram, and W. F. Libby
1949   The half-life of radiocarbon (C^{14}). *Physical Review 75:1825–1833.*
Engstrand, L. G.
1965   Stockholm natural radiocarbon measurements VI. *Radiocarbon 7:257–290.*
Erlenkeuser, H.
1979   A thermal diffusion plant for radiocarbon isotope enrichment from natural samples. In *Radiocarbon dating,* edited by R. Berger and H. E. Suess. Berkeley: University of California Press. Pp. 216–237.
Ertel, J. R., J. E. Hedges, and E. M. Perdue
1984   Lignin signature of aquatic humic substances. *Science 223:485–487.*
Evin, J., B. Kromer, H. Schoch-Fischer, M. Bruns, M. Munnich, D. Berdau, J. C.Vogel, and K. O. Munnich
1985   25 years of tropospheric ^{14}C observation in central Europe. *Radiocarbon 27:1–19.*

Evin, J.
1983   Materials of terrestrial origin used for radiocarbon dating. In *¹⁴C and archaeology*, edited by W. G. Mook and H. T. Waterbolk. [*PACT 8*:233–275]. Strasbourg: Council of Europe. Pp. 233–275.

Evin, J., J. Marechal, and G. Marien
1983   Lyon natural radiocarbon measurements IX. *Radiocarbon 25*:59–128.
1985   Lyon natural radiocarbon measurements X. *Radiocarbon 27*:386–454.

Evin, J., J. Marechal, C. Pachiaudi, and J. J. Puissegur
1980   Conditions involved in dating terrestrial shells. *Radiocarbon 22*:545–555.

Ewer, D. W.
1971   Thoughts on radiocarbon dating. *Antiquity 45*:201–202.

Fairhall, A. W., W. R. Shell, and Y. Takashima
1961   Apparatus for methane synthesis for radiocarbon dating. *Review of Scientific Instruments 32*:323.

Fairhall, A. W., and J. A. Young
1970   Radiocarbon in the environment. In *Radionuclides in the environment*, edited by E. C. Freiling. Washington, D.C.: American Chemical Society. Pp. 401–418.

Fairhall, A. W., and J. A. Young
1985   Historical ¹⁴C measurements from the Atlantic, Pacific, and Indian Oceans. *Radiocarbon 27*:473–507.

Fakid, A. F., M. A. F. El-Daoushy, I. U. Olsson, and F. H. Oro
1978   The EDTA and HCl methods of pre-testing bones. *Geologiska Föreningens i Stockholm Förhandlingar 100*:213–219.

Ferguson, C. W.
1968   Bristlecone pine: Science and esthetics. *Science 159*:839–846.
1979   Dendrochronology of bristlecone pine, *Pinus longaeva*. *Environment International 2*:209–214.

Ferguson, C. W., and D. A. Graybill
1983   Dendrochronopology of bristlecone pine: A progress report. *Radiocarbon 25*:287–288.

Ferguson, C. W., B. Huber, and H. E. Suess
1966   Determination of the age of Swiss lake dwellings as an example of dendrochronologically-calibrated radiocarbon dating. *Zeitschrift für Naturforschung 21A*:1173–1177.

Fergusson, G. J.
1955   Radiocarbon dating system. *Nucleonics 13*:18–23.

Fergusson, G. J. and W. F. Libby
1963   UCLA radiocarbon dates II. *Radiocarbon 5*:1–22.

Fergusson, G. J. and T. A. Rafter
1953   New Zealand ¹⁴C age measurements I. *New Zealand Journal of Science and Technology*, Section B. *35*:127–128.

Fishman, B., H. Forbes, and B. Lawn
1977   University of Pennsylvania radiocarbon dates XIX. *Radiocarbon 19*:188–228.

Fleming, S.
1976   Radiocarbon dating. In *Dating in archaeology: A guide to scientific techniques*. New York: St. Martin's Press. Chap. 3.

Fletcher, J. M.
1970   Radiocarbon dating of medieval timber-framed cruck cottages. In *Scientific methods in medieval archaeology*, edited by R. Berger. Berkeley: University of California Press. Pp. 141–166.

Flint, R. F., and E. S. Deevey, Jr.
1961   Editorial statement. *Radiocarbon 3:*ix–x.
1962   Editorial statement. *Radiocarbon 4:*v–vi.
Folk, R. L., and S. Valastro
1976   Successful technique for dating of lime mortar by carbon-14. *Journal of Field Archaeology 3:*203–208.
Fraser, I., H. A. Polach, R. B. Temple, and R. Gillespie
1974   Purity of benezene synthesized for liquid scintillation ¹⁴C dating. In *Liquid scintillation counting: Recent developments,* edited by P. E. Stanley and B. A. Scoggins. New York: Academic Press. Pp. 173–182.
Freundlich, J. C.
1973   Natural radon as a source of low level laboratory contamination. In *Proceedings of the Eighth International Radiocarbon Dating Conference,* compiled by T. A. Rafter and T. Grant-Taylor. Wellington: Royal Society of New Zealand. Pp. 537–546.
Freundlich, J. C., and B. Schmidt
1983   Calibrated ¹⁴C dates in central Europe. *Radiocarbon 25:*279–286.
Friedlander, G., J. W. Kennedy, and J. M. Miller
1964   *Nuclear and radiochemistry* (2nd ed.). New York: John Wiley and Sons.
Fritts, H. C.
1969   Bristlecone pine in the White Mountains of California: Growth and ring-width characteristics. *Papers of the Laboratory of Tree-Ring Research* No. 4. Tucson: University of Arizona.
Geyh, M. A.
1965   Proportional counter equipment for sample dating with ages exceeding 60,000 years B.P. without enrichment. In *Proceedings of the Sixth International Conference, Radiocarbon and Tritium Dating,* compiled by R. M. Chatter and E. A. Olson. Springfield, Virginia: Clearinghouse for Federal Scientific and Technical Information. Pp. 29–36.
Geyh, M. A., G. Roeschmann, T. A. Wijmstra, and A. A. Middeldorp
1983   The unreliability of ¹⁴C dates obtained from buried sandy podzols. *Radiocarbon 25:*409–416.
Geyh, M. A., and H. Schneekloth
1964   Hannover radiocarbon measurements III. *Radiocarbon 6:*251–268.
Gilet-Blein, N., G. Marien, and J. Evin
1980   Unreliability of ¹⁴C dates from organic matter of soils. *Radiocarbon 22:*919–929.
Gillespie, R.
1984   *Radiocarbon User's Handbook.* Oxford: Oxonian Rewley Press.
Gillespie, R., J. A. J. Gowlett, and R. E. M. Hedges
1984a   Recent development in archaeological dating using an accelerator. *Nuclear Instruments and Methods in Physics Research 233*(B5):308–311.
Gillespie, R., R. E. M. Hedges, and J. O. Wand
1984b   Radiocarbon dating of bone by accelerator mass spectrometry. *Journal of Archaeological Science 11:*165–170.
Gillespie, R., J. A. J. Gowlett, E. G. Hall, and R. E. M. Hedges
1984c   Radiocarbon measurements by accelerator mass spectrometry: An early selection of dates. *Archaeometry 26:*15–20.
Gillespie, R., J. A. J. Gowlett, E. T. Hall, R. E. M. Hedges, and C. Perry
1985   Radiocarbon dates from the Oxford AMS system: Archaeometry datelist 2. *Archaeometry 27:*237–246.

Gillespie, R., and H. A. Polach
    1979   The suitability of marine shells for radiocarbon dating of Australian prehistory.
              In *Radiocarbon dating*, edited by R. Berger and H. E. Suess. Berkeley: Uni-
              versity of California Press. Pp. 404–421.

Gittins, G. O.
    1984   Radiocarbon chronometry and archaeological thought. Unpublished Ph.D.
              dissertation, University of California, Los Angeles.

Godwin, H.
    1954   Carbon-14 Dating Symposium in Copenhagen. *Nature (London) 174:*868.
    1959   Carbon-Dating Conference at Groningen. *Nature (London) 184:*1365–1366.

Godwin, H.
    1962a   Radiocarbon Dating Fifth International Conference. *Nature (London) 195:*943–
              945.
    1962b   Half-life of radiocarbon. *Nature (London) 195:*984.

Goh, K. M., and B. P. J. Molloy
    1973   Reliability of radiocarbon dates from buried charcoals. In *Proceedings of the
              Eighth International Radiocarbon Dating Conference*, compiled by T. A. Rafter
              and T. Grant-Taylor. Wellington: Royal Society of New Zealand. Pp. 565–
              581.

Goh, K. M., B. P. J. Molloy, and T. A. Rafter
    1977   Radiocarbon dating of Quaternary loess deposits, Banks Peninsula, Canter-
              bury, New Zealand. *Quaternary Research 7:*177–196.

Goodfriend, G. A., and D. G. Hood
    1983   Carbon isotope analysis of land snail shells: Implications for carbon sources
              and radiocarbon dating. *Radiocarbon 25:*810–830.

Goslar, T., and M. F. Pazdur
    1985   Contamination studies on mollusk shell samples. *Radiocarbon 27:*33–42.

Gove, H. E.
    1981   Ultrasensitive mass spectrometry with a tandem Van de Graaff accelerator.
              In *Symposium on accelerator mass spectrometry*, edited by W. Kutshera.
              Springfield, Virginia: National Technical Information Services. Pp. 16–22.

Gove, H. E. (editor)
    1978   *Proceedings of the First Conference on Radiocarbon Dating with Accelerators.*
              Rochester, New York: University of Rochester.

Gowlett, J. A. J., E. T. Hall, R. E. M. Hedges, and C. Perry
    1986   Radiocarbon dates from the Oxford AMS system: Archaeometry datelist 3.
              *Archaeometry 28:*116–125.

Grant-Taylor, T. L.
    1973   Conditions for the use of calcium carbonate as a dating material. In *Proceedings
              of the Eighth International Radiocarbon Conference*, compiled by T. A. Rafter
              and T. Grant-Taylor. Wellington: Royal Society of New Zealand. Pp. 592–
              595.

Grey, D. C., and P. E. Damon
    1970   Sunspots and radiocarbon dating in the Middle Ages. In *Scientific methods in
              medieval archaeology*, edited by R. Berger. Berkeley: University of California
              Press. Pp. 167–182.

Griffin, J. B.
    1949 C^{14} dates. *American Antiquity 15:*80.
    1965   Radiocarbon dating and the cultural sequence in the eastern United States. In
              Proceedings of the Sixth International Conference Radiocarbon and Tritium

Dating, compiled by R. M. Chatters and E. A. Olson. Springfield, Virginia: Clearinghouse for Federal Scientific and Technical Information. Pp. 117–130.

Grootes, P. M., W. G. Mook, J. C. Vogel, A. E. de Vries, A. Haring, and J. Kistmaker
1975   Enrichment of radiocarbon for dating samples up to 75,000 years. *Zeitschrift füer Naturforschung 30A:*1–14.

Grosse, A. V.
1934   An unknown radioactivity. *Journal of the American Chemical Society 56:*1922.

Gulliksen, S.
1980   Isotopic fractionation of Norwegian materials for radiocarbon dating. *Radiocarbon 22:*980–986.
1983   Radiocarbon database: A pilot project. *Radiocarbon 25:*661–666.

Gupta, S. K., and H. A. Polach
1985   *Radiocarbon dating practices at ANU.* Canberra: Radiocarbon Laboratory, Research School of Pacific Studies, Australian National University.

Haas, H.
1979   Specific problems with liquid scintillation counting of small benzene volumes and background count rate estimation. In *Radiocarbon dating,* edited by R. Berger and H. E. Suess. Berkeley: University of California Press. Pp. 246–255.

Haas, H., and J. J. Banewics
1980   Radiocarbon dating of bone apatite using thermal release of CO_2. *Radiocarbon 22:*537–544.

Haas, H., V. Holliday, and R. Stuckenrath
1986   Dating of holocene stratigraphy with soluble and insoluble organic fractions at the Lubbock Lake archaeological site. Texas: an ideal case study. *Radiocarbon 28:*473–485.

Håkansson, S.
1979   Radiocarbon activity in submerged plants from various south Swedish lakes. In *Radiocarbon dating,* edited by R. Berger and H. E. Suess. Berkeley: University of California Press. Pp. 433–443.

Hall, E. T., and R. E. M. Hedges
1977   Carbon-14 dating of milligram samples by isotope-enriched mass spectrometry. *Abstracts, An International Symposium on Archaeometry and Archaeological Prospection.* University of Pennsylvania and University Museum, Philadelphia.

Harbottle, G., E. V. Sayre, and R. W. Stoenner
1979   Carbon 14 dating of small samples by proportional counting. *Science 206:*683–685.

Harkness, D. D.
1983   The extent of natural ^{14}C deficiency in the coastal environment of the United Kingdom. In *^{14}C and archaeology,* edited by W. G. Mook and H. T. Waterbolk. [*PACT 8:*355–364]. Strasbourg: Council of Europe. Pp. 352–364.

Harkness, D. D., and R. Burleigh
1974   Possible carbon-14 enrichment in high altitude wood. *Archaeometry 16:*121–127.

Hartley, P. E., and V. E. Church
1974   A low background liquid scintillation counter for ^{14}C. In *Liquid scintillation counting: Recent developments,* edited by P. E. Stanley and B. A. Scoggins. New York: Academic Press. Pp. 67–76.

Hassan, A. A.
1976   Geochemical and mineralogical studies on bone material and their implications

for radiocarbon dating. Unpublished Ph.D. dissertation, Southern Methodist University, Dallas.

Hassan, A. A., and P. E. Hare
1978   Amino acid analysis in radiocarbon dating of bone collagen. In *Archaeological chemistry II*, edited by G. F. Carter. Washington, D.C.: American Chemical Society. Pp. 109–116.

Hassan, A. A., and D. J. Ortner
1977   Inclusions in bone materials as a source of error in radiocarbon dating. *Archaeometry 19:*131–135.

Hassan, A. A., J. D. Termine, and C. V. Haynes
1977   Mineralogical studies on bone apatite and their implications for radiocarbon dating. *Radiocarbon 19:*364–384.

Hayes, F. N., D. L. Williams, and B. Rogers
1953   Liquid scintillation counting of natural C-14. *Physical Review 92:*512–513.

Haynes, C. V., Jr.
1965   Carbon-14 dates and early man in the new world. In *Proceedings of the Sixth International Conference on Radiocarbon and Tritium Dating*, compiled by R. M. Chatters and E. A. Olson. Springfield, Virginia: Clearinghouse for Federal Scientific and Technical Information. Pp. 145–164.

1966   Radiocarbon samples: Chemical removal of plant contaminants. *Science 151:*1391–1392.

1967   Bone organic matter and radiocarbon dating. In *Radiocarbon dating and methods of low-level counting*. Vienna: International Atomic Energy Agency. Pp. 163–168.

1968   Radiocarbon: Analysis of inorganic carbon of fossil bone and enamel. *Science 161:*687–688.

1978   Dating of archaeological and geological sites. In *Proceedings of the First Conference on Radiocarbon Dating with Accelerators*, edited by H. E. Gove. Rochester, New York: University of Rochester. Pp. 276–288.

1982   Were Clovis progenitors in Beringia? In *Paleoecology of Beringia*, edited by D. M. Hopkin, J. V. Matthews, Jr., C. E. Schweger, and S. B. Young. New York: Academic Press. Pp. 383–398.

Haynes, C. V., Jr., A. R. Doberenz, and J. A. Allen
1966   Geological and geochemical evidence concerning the antiquity of bone tools from Tule Springs, site 2, Clark County, Nevada. *American Antiquity 31:*517–521.

Haynes, V. and G. Agogino
1960   Geological significance of a new radiocarbon date from the Lindenmeir site. *The Denver Museum of Natural History Proceedings*, No. 9. Denver: Denver Museum of Natural History.

Hedges, R. E. M.
1981   Radiocarbon dating with an accelerator: Review and preview. *Archaeometry 23:*3–18.

1983 ^{14}C dating by the accelerator technique. In *^{14}C and archaeology*, edited by W. G. Mook and H. T. Waterbolk. [*PACT 8:*165–175]. Strasbourg: Council of Europe. Pp. 165–175.

Hedges, R. E. M., and J. A. J. Gowlett
1986   Radiocarbon dating by accelerator mass spectrometry. *Scientific American 254:*100–107.

Hedges, R. E. M. and C. B. Moore
1978   Enrichment of ^{14}C and radiocarbon dating. *Nature 276:*255–257.

References

185

Hedges, R. E. M., P. Ho, and C. B. Moore
  1980   Enrichment of carbon-14 by selective lasar photolysis of formaldehyde. *Applied Physics 23*:25–32.
Heinemeier, J., and H. H. Anderson
  1983   Production of C⁻ directly from CO_2 using the Anis sputter source. *Radiocarbon 25*:761–769.
Ho, T. Y., L. F. Marcus, and R. Berger
  1969   Radiocarbon dating of petroleum-impregnated bone from tar pits at Rancho La Brea, California. *Science 164*:1051–1052.
Holliday, V. T. and E. Johnson
  1986   Re-evaluation of the first radiocarbon age for the Folsom culture. *American Antiquity 51*:332–338.
Horn, W.
  1970   The potential and limitations of radiocarbon dating in the Middle Ages: The art historian's view. In *Scientific methods in medieval archaeology*, edited by R. Berger. Berkeley: University of California Press. Pp. 23–87.
Horrocks, D. L.
  1976   The mechanisms of the liquid scintillation process. In *Liquid scintillation science and technology*, edited by A. A. Noujaim, C. Edies, and L. E. Weibe. New York: Academic Press. Pp. 1–16.
Horvatinčić, N., D. Srdoč, B. Obelič, and A. Sliepcevič
  1983   Radiocarbon dating of fossil bones: development of a new technique for sample processing. In *¹⁴C and archaeology*, edited by W. G. Mook and H. T. Waterbolk. [PACT 8:377–384]. Strasbourg: Council of Europe. Pp. 377–384.
Hubbs, C. L., G. S. Bien, and H. E. Suess
  1960   La Jolla natural radiocarbon measurements I. *Radiocarbon 2*:197–223.
Hughes, E. E., and Mann, W. B.
  1964   The half-life of carbon-14: Comments on the mass-spectrometric method. *International Journal of Applied Radiation Isotopes 15*:97–100.
International Atomic Energy Agency
  1967   *Radioactive dating and methods of low-level counting*. Vienna: International Atomic Energy Agency.
International Study Group
  1982   An inter-laboratory comparison of radiocarbon measurements in tree rings. *Nature (London) 298*:619–623.
Irving, W. N., and C. R. Harrington
  1973   Upper Pleistocene radiocarbon-dated artifacts from the Northern Yukon. *Science 179*:335–340.
Jansen, H. S.
  1970   Secular variations of radiocarbon in New Zealand and Australian trees. In *Radiocarbon variations and absolute chronology*, edited by I. U. Olsson. Stockholm: Almqvist & Wiksell. Pp. 262–274.
Jeffreys, D., D. Larson, and J. D. French
  1972   C¹⁴ dating with nuclear track emulsions. *American Journal of Physics 40*:1400–1402.
Jelinek, A. J.
  1962   An index of radiocarbon dates associated with cultural materials. *Current Anthropology 3*:451–477.
Johnson, F.
  1942   The Boylston Street Fishweir. *Papers of the Robert S. Peabody Foundation for Archaeology 2*:1–211.

1949   The Boylston Street Fishweir II. *Papers of the Robert S. Peabody Foundation for Archaeology 4*:1–133.

1951   Introduction. In *Radiocarbon dating. Memoirs of the Society for American Archaeology 8*:1–3 [*American Antiquity 17*(1, Part 2):1–3].

1952   The significance of the dates for archaeology and geology. In W. F. Libby, *Radiocarbon dating*. Chicago: University of Chicago Press. Pp. 97–111.

1955   Reflections upon the significance of radiocarbon dates. In W. F. Libby, *Radiocarbon dating* (2nd ed.). Chicago: University of Chicago Press. Pp. 141–161.

1959   A bibliography of radiocarbon dating. *Radiocarbon 1*:199–214.

1965   The impact of radiocarbon dating upon archaeology. In *Proceedings of the Sixth International Conference Radiocarbon and Tritium Dating*, compiled by R. M. Chatters and E. A. Olson. Springfield, Virginia: Clearinghouse for Federal Scientific and Technical Information. Pp. 762–780.

Johnson, F., Arnold, J. R., and R. F. Flint
1957   Radiocarbon dating. *Science 125*:240–242.

Johnson, F., F. Rainey, D. Collier, and R. F. Flint
1951   Radiocarbon dating, a summary. In *Radiocarbon dating, Memoirs of the Society for American Archaeology 8*:59–63. [*American Antiquity 17*(1, part 2):59–63].

de Jong, A. F. M., and W. G. Mook
1980   Medium-term atmospheric ^{14}C variations. *Radiocarbon 22*:267–272.

de Jong, A. F. M., W. G. Mook, and B. Becker
1979   Confirmation of Suess wiggles: 3200–3700 B. C. *Nature (London) 280*:48–49.

Kamen, M. D.
1963   Early history of carbon-14. *Science 140*:584–590.

1985   *Radiant science, dark politics, a memoir of the nuclear age*. Berkeley: University of California Press.

Karrow, P. E., B. G. Warner, and P. Fritz
1984   Corry Bog, Pennsylvania: A case study of the radiocarbon dating of marl. *Quaternary Research 21*:326–336.

Kaufman, T. S.
1980   Early prehistory of the Clear Lake area, Lake County, California. Unpublished Ph.D. dissertation, University of California, Los Angeles.

Keisch, B.
1972   *Secrets of the past: Nuclear energy application in art and archaeology*. Washington, D.C.: U.S. Atomic Energy Commission.

Keith, M. L., and G. B. Anderson
1963   Radiocarbon dating: Fictitious results with mollusk shells. *Science 141*:634–637.

Kelley, D. H.
1983   The Maya calendar correlation problem. In *Civilization in the ancient Americas, essays in honor of Gordon R. Willey*, edited by R. M. Leventhal and A. L. Lolata. Albuquerque: University of New Mexico Press. Pp. 157–186.

Kemp, B.
1980   Egyptian radiocarbon dating: A reply to James Mellaart. *Antiquity 54*:25–28.

Kigoshi, K., N. Suzuki, and M. Shiraki
1980   Soil dating by fractional extraction of humic acid. *Radiocarbon 22*:853–857.

Klein, J., J. C. Lerman, P. E. Damon, and T. Lincik
1980   Radiocarbon concentration in the atmosphere: 8000 year record of variation in tree-rings. *Radiocarbon 22*:950–961.

Klein, J., J. C. Lerman, P. E. Damon, and E. K. Ralph
1982   Calibration of radiocarbon dates: Tables based on the consensus data of the workshop on calibrating the radiocarbon time scale. *Radiocarbon 24*:103–150.

Korff, S. A.
1940   On the contribution to the ionization at sea-level produced by the neutrons in the cosmic radiation. *Terrestrial Magnetism and Atmospheric Electricity 45:*133–134.

Korff, S. A., and W. E. Danforth
1939   Neutron measurements with boron-trifluoride counters. *Physical Review 55:*980.

Kovar, A. J.
1966   Problems in radiocarbon dating at Teotihuacan. *American Antiquity 31:*427–430.

Kruse, H. H., T. W. Linick, and H. E. Suess
1980   Computer-matched radiocarbon dates of floating tree-ring series. *Radiocarbon 22:*260–266.

Kulp, J. L.
1954a   Errors and limitations of the screenwall technique. In Conference on Radiocarbon Dating. R. S. Peabody Foundation, Phillips Academy, Andover, Massachusetts. Pp. 136–143. Unpublished transcript of meeting.
1954b   Assumptions of cosmic ray flux. In Conference on Radiocarbon Dating. R. S. Peabody Foundation, Phillips Academy, Andover, Massachusetts. Pp. 50–55. Unpublished transcript of meeting.

Kulp, J. L., H. W. Freely, and L. E. Tryon
1951   Lamont natural radiocarbon measurements, I. *Science 114:*565–568.

Kulp, J. L., L. E. Tryon, W. R. Eckelman, and W. A. Snell
1952   Lamont natural radiocarbon measurements, II. *Science 116:*409–414.

Kurie, F. N. D.
1934   A new mode of disintegration induced by neutrons. *Physical Review 45:*904–905.

Kutschera, W. (editor)
1981   *Symposium on accelerator mass spectrometry.* Argonne: Argonne National Laboratory.

Kutschera, W.
1983   Accelerator mass spectrometry: From nuclear physics to dating. *Radiocarbon 25:*677–691.

Labeyrie, J., and G. Delibrias
1964   Dating of old mortars by the carbon-14 method. *Nature (London) 201:*742.

LaMarche, V. C. and T. P. Harlan
1973   Accuracy of tree ring dating of bristlecone pine for calibration of radiocarbon time scale. *Journal of Geophysical Research 78:*8849–8858.

Leavitt, S. W., and A. Long
1982   Evidence for $^{13}C/^{12}C$ fractionation between tree leaves and wood. *Nature (London) 298:*742–743.

Lee, R. E.
1981   Radiocarbon: Ages in error. *Anthropological Journal of Canada 19:*9–29.

Lerman, J. C., W. G. Mook, and J. C. Vogal
1970   C14 in tree rings from different localities. In *Radiocarbon variations and absolute chronology,* edited by I. U. Olsson. Stockholm: Almqvist & Wiksell. Pp. 275–299.

Levi, H.
1955a   Bibliography of radiocarbon dating. *Quaternaria 2:*257–263.
1955b   Radiocarbon dating conference in Cambridge. *Nature (London) 176:*727–728.
1957   Bibliography of radiocarbon dating. *Quaternaria 4:*205–210.

Levin, I., B. Kromer, H. Schoch–Fisher, M. Bruns, M. Munnich, D. Berdau, J. C. Vogel, and K. O. Munnich

1985   25 years of tropospheric ^{14}C observations in central Europe. *Radiocarbon 27*:1–
            19.
Libby, L. M.
    1973   Globally stored organic carbon and radiocarbon dates. *Journal of Geophysical
            Research 78*:7667–7670.
    1981   Willard Frank Libby 1908–1980. In *Radiocarbon and tritium, Willard F. Libby
            collected papers* (Vol. 1). Santa Monica: Geo Science Analytical. Pp. 5–6.
    n.d.   *The isotope people.* Unpublished manuscript.
Libby, L. M., and W. F. Libby
    1973   Vulcanism and radiocarbon dates. In *Proceedings of the Eighth International
            Radiocarbon Conference,* compiled by T. A. Rafter and T. Grant-Taylor. Wel-
            lington: Royal Society of New Zealand. Pp. 86–89.
Libby, L. M., and H. R. Lukens
    1973   Production of radiocarbon in tree rings by lightening bolts. *Journal of Geo-
            physical Research 78*:5902.
Libby, W. F.
    1932   Simple amplifier for Geiger–Muller counters. *Physical Review 42*:440–441.
    1933   Radioactivity of ordinary elements, especially Samarium and Neodymium:
            method of detection. Unpublished Ph.D. dissertation, University of California,
            Berkeley.
    1934   Radioactivity of Neodynium and Samarium. *Physical Review 46*:196.
    1946   Atmospheric Helium three and radiocarbon from cosmic radiation. *Physical
            Review 69*:671–672.
    1947   Measurement of radioactive tracers particularly C^{14}, S^{35}, T and other longer-
            lived low-energy activities. *Chemistry 19*:2–6.
    1951   Radiocarbon dates, II. *Science 114*:291–296.
    1952a   *Radiocarbon dating.* Chicago: University of Chicago Press.
    1952b   Chicago radiocarbon dates, III. *Science 116*:673–681.
    1954a   Chicago radiocarbon dates IV. *Science 119*:135–140.
    1954b   Chicago radiocarbon dates V. *Science 120*:733–742.
    1955   *Radiocarbon dating* (2nd ed.). Chicago: University of Chicago Press.
    1961a   Radiocarbon dating [Nobel Lecture, December 12, 1960]. *Les Prix Nobel En
            1960.* Stockholm: Nobel Foundation. Pp. 95–112.
    1961b   Radiocarbon dating. *Science 133*:621–629.
    1963   Accuracy of radiocarbon dates. *Science 140*:278–280 [*Antiquity 37*:213–218].
    1964a   Thirty years of atomic chemistry. *Annual Review of Physical Chemistry 15*:1–
            6.
    1964b   Berkeley radiochemistry. *Annual Review of Physical Chemistry 15*:7–12.
    1965a   *Radiocarbon dating* (2nd ed., 1st Phoenix ed.). Chicago: University of Chicago
            Press.
    1965b   Natural radiocarbon and tritium in retrospect and prospect. In *Proceedings of
            the Sixth International Conference on Radiocarbon and Tritium Dating,* com-
            piled by R. M. Chatters and E. A. Olson. Springfield, Virginia: Clearinghouse
            for Federal Scientific and Technical Information. Pp. 745–751.
    1967   History of radiocarbon dating. In *Radioactive dating and methods of low level
            counting.* Vienna: International Atomic Energy Agency. Pp. 3–25.
    1970a   Radiocarbon dating. *Philosophical Transactions of the Royal Society of London
            269A*:1–10.
    1970b   Ruminations on radiocarbon dating. In *Radiocarbon variations and absolute
            chronology,* edited by I. U. Olsson. Stockholm: Almqvist & Wiksell. Pp. 629–
            640.

1973 Radiocarbon dating, memories and hopes. In *Proceedings of the Eighth International Radiocarbon Conference*, compiled by T. A. Rafter and T. Grant-Taylor. Wellington: Royal Society of New Zealand. Pp. xxvii–xliii.

1979a Interview with W. F. Libby, April 12, 1979, on file at the Center for the History of Physics, American Institute of Physics.

1979b Radiocarbon dating in the future: Thirty years after inception. *Environment International 2*:205–207.

1980 Archaeology and radiocarbon dating. *Radiocarbon 22*:1017–1020.

1982 Nuclear dating: An historical perspective. In *Nuclear and Chemical dating techniques interpreting the environmental record*, edited by L. A. Currie. Washington, D.C.: American Chemical Society. Pp. 1–4.

Libby, W. F., E. C. Anderson, J. R. Arnold
1949 Age determination by radiocarbon content: World wide assay of natural radiocarbons. *Science 109*:227–228.

Libby, W. F., and D. D. Lee
1939 Energies of the soft beta-radiations of Rubidium and other bodies. Method for their determination. *Physical Review 55*:245.

Lingenfelter, R. E.
1963 Production of Carbon 14 by cosmic-ray neutrons. *Review of Geophysics 1*:35–55.

Linick, T. W., H. E. Suess, and B. Becker
1985 La Jolla measurements of radiocarbon in South German oak tree-ring chronologies. *Radiocarbon 27*:20–32.

Litherland, A. E.
1980 Ultrasensitive mass spectrometry with accelerators. *Annual Review of Nuclear Particle Science 30*:437–473.

1984 Accelerator mass spectrometry. *Nuclear Instruments and Methods in Physics Research 233*(B5):100–108.

Long, A.
1965 Techniques of methane production for C^{14} dating. In *Proceedings of the Sixth International Conference Radiocarbon and Tritium Dating*, compiled by R. M. Chatters and E. A. Olson, Compilers. Springfield Virginia: Clearinghouse for Federal Scientific and Technical Information. Pp. 37–40.

Long, A., R. B. Hendershott, and P. S. Martin
1983 Radiocarbon dating of fossil eggshell. *Radiocarbon 25*:533–539.

Long, A., and A. B. Muller
1981 Arizona radiocarbon dates X. *Radiocarbon 23*:191–217.

Long, A., and B. Rippeteau
1974 Testing contemporaneity and averaging radiocarbon dates. *American Antiquity 39*:205–215.

Longin, R.
1971 New method of collagen extraction for radiocarbon dating. *Nature (London) 230*:241–242.

Lowdon, J. A., and W. Blake, Jr.
1968 Geological survey of Canada radiocarbon dates VII. *Radiocarbon 10*:207–245.

1970 Geological Survey of Canada radiocarbon dates IX. *Radiocarbon 12*:46–86.

Lowdon, J. A., J. G. Fyles and W. Blake, Jr.
1967 Geological survey of Canada radiocarbon dates VI. *Radiocarbon 9*:156–197.

MacKie, E., J. Collis, D. W. Ewer, A. Smith, H. Suess, and C. Renfrew
1971 Thoughts on radiocarbon dating. *Antiquity 45*:197–204.

Mangerud, J., and S. Gulliksen
  1975   Apparent radiocarbon ages of recent marine shells from Norway, Spitsbergen,
            and Arctic Canada. *Quaternary Research 5:*263–273.
Mann, J. C., and D. K. Messman
  1976   Dating archaeological material. *Current Anthropology 17:*484–485.
Mann, W. B.
  1983   An international reference materials for radiocarbon dating. *Radiocarbon
            25:*519–527.
Marlowe, G.
  1980   W. F. Libby and the archaeologists, 1946–1948. *Radiocarbon 22:*1005–1014.
Mast, T. S., and Muller, R. A.
  1980   Radioisotope detection and dating with accelerators. *Nuclear Science Appli-
            cations 1:*7–32.
Masters, P. M. and J. L. Bada
  1979   Amino acid racemization dating of fossil shell from southern California. In
            Radiocarbon Dating, edited by R. Berger and H. E. Suess. Berkeley: University
            of California Press, Pp. 757–776.
Matson, F. R.
  1955   Charcoal concentration from early sites for radiocarbon dating. *American An-
            tiquity 21:*162–169.
McAulay, I. R., and W. A. Watts
  1961   Dublin radiocarbon dates I. *Radiocarbon 3:*26–38.
Mc Burney, C. B. M.
  1952   Radiocarbon dating results from the Old World. *Antiquity 26:*35–40.
McCallum, K. J.
  1955   Carbon-14 age determinations at the University of Saskatchewan (I). *Trans-
            actions of the Royal Society of Canada 49*(4):31–35.
McKerrell, H.
  1975   Correction procedures for C-14 dates. In *Radiocarbon: Calibration and pre-
            history,* edited by T. Watkins. Edinburgh: Edinburgh University Press. Pp.
            47–100.
McPhail, S. M.
  1982   Reliable radiocarbon dates from bones: The Sydney University experiment.
            In *Archaeometry: An Australasian perspective,* edited by W. Ambrose and P.
            Duerden. Canberra: Department of Prehistory, Research School of Pacific
            Studies, Australian National University. Pp. 336–342.
McPhail, S, M. Barbetti, R. Francey, T. Bird, and J. Dolezal
  1983 ^{14}C variations from Tasmanian trees—preliminary results. *Radiocarbon 25:*797–
            802.
Mellaart, J.
  1979   Egyptian and Near Eastern chronology: A dilemma? *Antiquity 53:*6–18.
Michael, H. N.
  1984   Extending the calibration of radiocarbon dates: The search for ancient wood.
            MASCA Journal 3(1):17–19.
Michels, J. W.
  1973   Radiocarbon dating. In *Dating Methods in Archaeology.* New York: Seminar
            Press. Chap. 9.
Middleton, R.
  1984   A review of ion sources for accelerator mass spectrometry. *Nuclear Instruments
            and Methods in Physics Research 233*(B5):193–199.

Moffett, J. C., and R. E. Webb
  1983   Database management systems, radiocarbon and archaeology. *Radiocarbon*
          25:667–668.
Molina-Cruz, A.
  1977   The relations of the southern trade winds to upwelling processes during the
          last 75,000 years. *Quaternary Research 8:*324–338.
Montgomery, C. G., and D. D. Montgomery
  1939   The intensity of neutrons of thermal energy in the atmosphere at sea level.
          *Physical Review 56:*10–12.
Mook, W. G.
  1983a   International comparison of proportional gas counters for ^{14}C activity mea-
          surements. *Radiocarbon 25:*475–484.
  1983b ^{x4}C calibration curves depending on sample time width. In *^{14}C and archaeology,*
          edited by W. G. Mook and H. T. Waterolk. Strasbourg: Council of Europe.
          Pp. 517–525.
  1984   Archaeological and geological interest in applying C^{14} AMS to small sam-
          ples. *Nuclear Instruments and Methods in Physics Research 233*(B5):297–
          302.
  1986   Business meeting, recommendations/resolutions adopted by the Twelfth In-
          ternational Radiocarbon Conference. *Radiocarbon 28:*799.
Mook, W. G., and H. J. Streurman
  1983   Physical and chemical aspects of radiocarbon dating. In *^{14}C and archaeology,*
          edited by W. G. Mook and H. T. Waterbolk. [PACT 8:31–54]. Strasbourg:
          Council of Europe. Pp. 31–54.
Morner, N. A.
  1977   The Gotenburg magnetic excursion. *Quaternary Research 7:*413–427.
Muller, R. A.
  1977   Radioisotope dating with a cyclotron. *Science 196:*489–494.
  1979   Radioisotope dating with accelerators. *Physics Today 32*(2):23–30.
Muller, R. A., E. J. Stephenson, and T. S. Mast
  1978   Radioisotope dating with an accelerator: A blind measurement. *Science 201:*347–
          348.
Munnich, K. O.
  1957   Heidelberg natural radiocarbon measurements I. *Science 126:*194–199.
Nelson, D. E., R. G. Korteling, and W. R. Scott
  1977   Carbon-14: Direct detection at natural concentrations. *Science 198:*507–
          508.
Nelson, D. E., R. E. Morlan, J. S. Vogel, J. R. Southon, and C. R. Harington
  1986   New radiocarbon dates on artifacts from the northern Yukon Territory: Hol-
          ocene not upper Pleistocene in age. *Science 232:*749–751.
Nelson, D. E., T. H. Loy, J. S. Vogel, and J. R. Southon
  1986   Radiocarbon dating blood residues on prehistoric stone tools. *Radiocarbon*
          28:170–174.
Neustupny, E.
  1970   The accuracy of radiocarbon dating. In *Radiocarbon variations and absolute
          chronology,* edited by I. U. Olsson. Stockholm: Almqvist & Wiksell. Pp. 23–
          34.
Nicholson, H. B.
  1977   The Pre-hispanic lowland Maya calendar. Unpublished paper, Department of
          Anthropology, University of California, Los Angeles.

Noakes, J. E.
1976   Considerations for achieving low level radioactivity measurements with liquid
           scintillation counters. In *Liquid scintillation counting,* edited by M. A. Crook
           and P. Johnson. (Vol. 4). London: Heyden. Pp. 189–204.
Noakes, J. E., S. M. Kim, and J. J. Stipp
1965   Chemical and counting advances in liquid scintillation age dating. In *Proceedings
           of the Sixth International Conference Radiocarbon and Tritium Dating,* com-
           piled by R. M. Chatters and E. A. Olson. Springfield, Virginia: Clearinghouse
           for Federal Scientific and Technical Information. Pp. 68–92.
Noakes, J. E., M. P. Neary, and J. D. Spaulding
1974   A new liquid scintillation counter for measurement of trace amounts of ^{3}H and
           ^{14}C In *Liquid scintillation counting: Recent developments,* edited by P. E.
           Stanley and B. A. Scoggins. New York: Academic Press. Pp. 53–66.
Nobel Foundation
1964   *Nobel Lectures, Chemistry 1942–1962.* Amsterdam: Elsevier.
Nydal, R.
1983a   Optimal number of samples and accuracy in dating problems. In *^{14}C and ar-
           chaeology,* edited by W. G. Mook and H. T. Waterbolk. [*PACT 8*:107–122].
           Strasbourg: Council of Europe. Pp. 107–122.
1983b   The radon problem in ^{14}C dating. *Radiocarbon 25*:501–510.
Nydal, R., S. Gulliksen, K. Lovseth, and F. Skogseth
1985   Trondheim natural radiocarbon measurements IX. *Radiocarbon 27*:525–609.
Oakley, K. P.
1966   *Frameworks for dating fossil man* (2nd ed.). Chicago: Aldine.
Oeschger, H., J. Houtermans, H. Loosli, and M. Wahlen
1970   The constancy of cosmic radiation from isotope studies in meteorites and on
           the earth. In *Radiocarbon variations and absolute chronology,* edited by I.
           U. Olsson. Stockholm: Almqvist & Wiksell. Pp. 471–496.
Oeschger, H., U. Siegenthaler, U. Schotterer, and A. Gugelmann
1975   A box diffusion model to study the carbon dioxide exchange in nature. *Tellus
           27*:168–192.
Ogden, J. G.
1977   The use and abuse of radiocarbon dating. *Annals of the New York Academy
           of Science 288*:167–188.
Ogden, J. G., and R. J. Hay
1965   Ohio Wesleyan University natural radiocarbon measurements II. *Radiocarbon
           7*:166–173.
1973   Ohio Wesleyan University natural radiocarbon measurements V. *Radiocarbon
           15*:350–366.
Olson, E. A.
1963   The problem of sample contamination in radiocarbon dating. Unpublished Ph.D.
           dissertation, Columbia University.
Olson, E. A., and W. S. Broecker
1959   Lamont natural radiocarbon measurements V. *American Journal of Science
           Radiocarbon Supplement 1*:1–28.
1961   Lamont natural radiocarbon measurements VII. *Radiocarbon 3*:141–175.
Olsson, I. U.
1958   A C^{14} dating station using CO_2 proportional counting. *Arkiv for Fysik 13*:37.
1968   Modern aspects of radiocarbon datings. *Earth-Science Reviews 4*:203–218.
1970a   (editor) *Radiocarbon variations and absolute chronology.* Stockholm: Almqvist
           & Wiksell.

1970b	The use of oxalic acid as a standard. In *Radiocarbon variations and absolute chronology*, edited by I. U. Olsson. Stockholm: Almqvist & Wiksell. P. 17.
1974a	Some problems in connection with the evaluation of C^{14} dates. *Geologiska Föreningens i Stockholm Förhandlingar 96*:311–320.
1974b	The Eighth International Conference on Radiocarbon Dating. *Geologiska Föreningens i Stockholm Förhandlingar 96*:37–44.
1979a	The radiocarbon contents of various reservoirs. In *Radiocarbon dating*, edited by R. Berger and H. E. Suess. Berkeley: University of California Press. Pp. 613–618.
1979b	The importance of the pretreatment of wood and charcoal samples. In *Radiocarbon dating*, edited by R. Berger and H. E. Suess. Berkeley: University of California Press. Pp. 135–146.
1980a	^{14}C in extractives from wood. *Radiocarbon 22*:515–524.
1980b	Content of ^{14}C in marine mammals from northern Europe. *Radiocarbon 22*:662–675.
1983a	Radiocarbon dating in the arctic region. *Radiocarbon 25*:393–394.
1983b	Dating non-terrestrial materials. In *^{14}C and archaeology*, edited by W. G. Mook and H. T. Waterbolk. [PACT 8:277–275]. Strasbourg: Council of Europe. Pp. 277–294.

Olsson, I. U., M. Klasson, and A. E. Abd El-Mageed
  1972 Uppsala natural radiocarbon measurements XI. *Radiocarbon 14*:247–271.
Olsson, I. U., and F. A. N. Osadebe
  1974 Carbon isotope variations and fractionation corrections in ^{14}C dating. *Boreas 3*:139–146.
Olsson, I. U., M. F. A. F. El-Daoushy, A. E. Abd El-Mageed, and M. Klasson
  1974 A comparison of different methods for pretreatment of bones I. *Geologiska föreningens i Stockholm Förhandlingar 96*:171–181.
Otlet, R. L.
  1979 An assessment of laboratory errors in liquid scintillation methods of ^{14}C dating. In *Radiocarbon dating*, edited by R. Berger and H. E. Suess. Berkeley: University of California Press. Pp. 256–267.
Otlet, R. L., and G. V. Evans
  1983 Progress in the application of miniature gas counter to radiocarbon dating of small samples. In *^{14}C and archaeology*, edited by W. G. Mook and H. T. Waterbolk. [*PACT 8*:57–70]. Strasbourg: Council of Europe. Pp. 57–70.
Otlet, R. L., G. Huxtable, G. V. Evens, D. G. Humphreys, T. D. Short, and S. J. Conchie
  1983 Development and operation of the Harwell small counter facility for the measurement of ^{14}C in very small samples. *Radiocarbon 25*:565–575.
Otlet, R. L., and A. J. Walker
  1983 The computer writing of radiocarbon reports and further developments in the storage and retrieval of archaeological data. In *^{14}C and archaeology*, edited by W. G. Mook and H. T. Waterbolk. [*PACT 8*:91–105]. Strasbourg: Council of Europe. Pp. 91–105.
Otlet, R. L., A. J. Walker, A. D. Hewson, and R. Burleigh
  1980 ^{14}C interlaboratory comparison in the UK: Experiment design, preparation and preliminary results. *Radiocarbon 22*:936–946.
Otlet, R. L., and R. M. Warchal
  1978 Liquid scintillation counting of low-level ^{14}C. In *Liquid scintillation counting*, edited by M. A. Crook and P. Johnson. (Vol. 5). London: Heyden. Pp. 210–216.

Panin, N., S. Panin, N. Herx, and J. E. Noakes
   1983   Radiocarbon dating of Danube delta deposits. *Quaternary Research 19:*249–255.

Pardi, R., and L. Marcus
   1977   Non-counting errors in ^{14}C dating. *Annals of the New York Academy of Sciences 288:*174–180.

Pavlish, L. A., and E. B. Banning
   1980   Revolutionary developments in carbon-14 dating. *American Antiquity 45:*290–297.

Payen, L. A., C. H. Rector, E. Ritter, R. E. Taylor, and J. E. Ericson
   1978   Comments on the Pleistocene age assignment and association of a human burial from the Yuha Desert, California. *American Antiquity 43:*448–453.

Payen, L. A., and R. E. Taylor
   1977   Man and Pleistocene fauna at Potter Creek Cave, California. *Journal of California Anthropology 3:*51–58.

Pearl, H. F.
   1963   A re-evaluation of time-variations in two geochemical parameters of importance in the accuracy of radiocarbon ages greater than four millennia. Unpublished M.A. thesis, Pacific Union College, Angwin, California.

Pearson, G. W.
   1979   Precise ^{14}C measurement by liquid scintillation counting. *Radiocarbon 21:*1–21.

Pearson, G. W., and M. G. L. Baillie
   1983   High-precision ^{14}C measurement of Irish oaks to show the natural atmospheric ^{14}C variations of the AD time period. *Radiocarbon 25:*187–196.

Pearson, G. W., J. R. Pilcher, M. G. Baillie, D. M. Corbett, and F. Qua
   1986   High-precision ^{14}C measurement of Irish Oaks to show the natural ^{14}C variations from AD 1840–5210 BC. *Radiocarbon 28:*911–934.

Pearson, G. W., J. R. Pilcher, M. G. L. Baillie, and J. Hillam
   1977   Absolute radiocarbon dating using a low altitude European tree-ring calibration. *Nature (London) 270:*25–28.

Phillips, P. J. Ford, and J. B. Griffin
   1951   Archaeological survey in the lower Mississippi alluvial valley 1940–1947. *Papers of the Peabody Museum of American Archaeology and Ethnology* No. 25.

Pilcher, J. R.
   1980   Radiocarbon calibration: Recent progress. In *Progress in scientific dating methods,* edited by R. Burleigh. Occasional Paper No. 21. London: British Museum. Pp. 45–51.

   1983   Radiocarbon calibration and dendrochronology—An introduction. In *Archaeology, dendrochronology, and the radiocarbon calibration curve,* edited by B. S. Ottaway. Occasional Paper No. 9, Department of Archaeology, University of Edinburgh. Pp. 5–14.

Pilcher, J. R., and M. G. L. Baillie
   1978   Implications of a European radiocarbon calibration. *Antiquity 52:*217–222.

Pilcher, J. R., M. G. L. Baillie, B. Schmidt, and B. Becker
   1984   A 7,272-year tree-ring chronology for western Europe. *Nature (London) 312:*150–152.

Polach, D.
   1980   The first 20 years of radiocarbon dating: an annotated bibliography, 1948–68; a pilot study. *Radiocarbon 22:*997–1004.

Polach, H. A.
1973   Cross checking of NBS oxalic acid and secondary laboratory radiocarbon dating standards. In *Proceedings of the 8th International Conference on Radiocarbon Dating*, compiled by T. A. Rafter and T. Grant-Taylor. Wellington: Royal Society of New Zealand. Pp. 688–717.
1974   Application of liquid scintillation spectrometers to radiocarbon dating. In *Liquid scintillation counting: Recent developments*, edited by P. E. Stanley and B. A. Scoggins. New York: Academic Press. Pp. 153–171.
1976   Radiocarbon dating as a research tool in archaeology—Hopes and limitations. In *Ancient Chinese bronzes and Southeast Asian metal and other archaeological artifacts*, edited by N. Barnard. Victoria: National Gallery of Victoria. Pp. 255–298.
1979   Correlations of ^{14}C activity of NBS oxalic acid with Arizona 1850 wood and ANU sucrose standards. In *Radiocarbon dating*, edited by R. Berger and H. E. Suess. Berkeley: University of California Press. Pp. 115–124.
1984   Radiocarbon targets for AMS: A review of perceptions, aims and achievements. *Nuclear Instruments and Methods in Physics Research 233*(B5): 259–264.
Polach, H. A., and J. Golson
1966   Collection of specimens for radiocarbon dating and interpretation of results, Manual No. 2, Australian Institute of Aboriginal Studies, Canberra.
Polach, H. A., J. Golson, J. F. Lovering, and J. J. Stipp
1968   ANU radiocarbon date list II. *Radiocarbon 10:*179–199.
Polach, H. A., J. Gower, H. Kojola, and A. Heinonen
1983   An ideal vial and cocktail for low-level scintillation sounting. In *Advances in scintillation counting*, edited by S. A. McQuarrie, C. Ediss, and L. I. Wiebe. Edmonton: University of Alberta Press. Pp. 508–525.
Polach, H., H. Kojola, J. Nurmi, and E. Soini
1984   Multiparameter liquid scintillation spectrometry. *Nuclear Instruments and Methods in Physics Research B5:*439–442.
Preece, R. C., R. Burleigh, M. P. Kerney, and E. A. Jarzembowski
1983   Radiocarbon age determinations of fossil *Magaritifera auricularia* (Spengler) from the River Thames in west London. *Journal of Archaeological Science 10:*249–257.
Pringle, R. W., W. Turchinetz, B. L. Funt, and S. S. Danyluk
1957   Radiocarbon age estimates obtained by an improved liquid scintillation technique. *Science 125:*69–70.
Purser, K. H., C. J. Russo, R. B. Liebert, H. E. Gove, D. Elmore, D. Ferro, A. E. Litherland, R. P. Beukens, K. H. Chang, L. R. Kilius, and H. W. Lee
1982   The application of electrostatic tandems to ultrasensitive mass spectrometry and nuclear dating. In *Nuclear and chemical dating techniques interpreting the environmental record*, edited by L. A. Currie. Washington: American Chemical Society. Pp. 45–74.
Quitta, H.
1967   The C14 chronology of the central and SE European neolithic. *Antiquity 41:*263–270.
Raaen, V. F., G. A. Ropp, and H. P. Raaen
1968   *Carbon-14*. New York: McGraw-Hill.
Radnell, C. J.
1980   The isotopic fractionation of ^{14}C and ^{13}C relative to ^{12}C. In *Proceedings of the 16th International Symposium on Archaeometry and Archaeological Pro-*

spection, Edinburgh, 1976, edited by E. A. Slater and J. O. Tate. Edinburgh: National Museum of Antiquities of Scotland. Pp. 360–392.

Radnell, C. J., M. J. Aitken, and R. L. Otlet
1979   *In situ* ^{14}C production in wood. In *Radiocarbon dating,* edited by R. Berger and H. E. Suess. Berkeley: University of California Press. Pp. 643–657.

Rafter, T. A.
1955a   Carbon dioxide as a substitute for solid carbon in ^{14}C age measurements. *New Zealand Journal of Science and Technology B36:*363–369.
1955b ^{14}C variations in nature and the effect on radiocarbon dating. *New Zealand Journal of Science and Technology B37:*20–38.

Rafter, T. A., and G. J. Fergusson
1957   The atom bomb effect. Recent increase in the ^{14}C content of the atmosphere, biosphere, and surface waters of the oceans. *New Zealand Journal of Science and Technology B38:*871–883.

Rafter, T. A., and T. Grant-Taylor (compliers)
1973   *Proceedings of the Eighth International Radiocarbon Dating Conference.* 2 vols. Wellington: Royal Society of New Zealand.

Ralph, E. K.
1955   University of Pennsylvania radiocarbon dates I. *Science 121:*149–151.
1959   University of Pennsylvania radiocarbon dates III. *American Journal of Science Radiocarbon Supplement 1:*45–58.
1965   Review of radiocarbon dates from Tikal and the Maya calendar correlation problem. *American Antiquity 30:*421–427.
1971   Carbon-14 dating. In *Dating techniques for the archaeologist,* edited by H. N. Michael and E. K. Ralph. Pp. 1–48. Cambridge, Massachusetts: MIT Press.

Ralph, E. K., and M. C. Han
1971   Potential of thermoluminescence dating. In *Science and archaeology,* edited by R. H. Brill. Cambridge, Massachusetts: MIT Press. Pp. 244–250.

Ralph, E. K., and H. N. Michael
1974   Twenty-five years of radiocarbon dating. *American Scientist 62:*553–560.

Ralph, E. K., H. N. Michael, and M. C. Han
1973   Radiocarbon dates and reality. *MASCA Newsletter 9:*1–20.

Rapkin, E.
1969   Development of the modern liquid scintillation counter. In *The current status of liquid scintillation counting,* edited by E. D. Bransome. New York: Grune & Stratton. Pp. 45–68.

Read, D. W.
1979   The effective use of radiocarbon dates in the seriation of archaeological sites. In *Radiocarbon dating,* edited by R. Berger and H. E. Suess. Berkeley: University of California Press. Pp. 89–94.

Renfrew, C.
1973   *Before civilization, the radiocarbon revolution and prehistoric Europe.* New York: Alfred A. Knopf.

Riggs, A. C.
1984   Major carbon-14 deficiency in modern snail shells from southern Nevada springs. *Science 224:*58–61.

Roberts, F. H. H.
1951   Radiocarbon dates and early man. In *Radiocarbon dating,* assembled by F. Johnson. *Society for American Archaeology, Memoirs 8:*20–21 [*American Antiquity 17*(1, Part 2)].

Robinson, S. W., and G. Thompson
1981   Radiocarbon corrections for marine shell dates with application to southern Pacific northwest coast prehistory. *Syesis 14:*45–57.
Robinson, S. W., and D. A. Trimble
1981   U. S. Geological Survey, Menlo Park, California radiocarbon measurements II. *Radiocarbon 23:*320–321.
Rogers, R. A., and L. D. Martin
1984   The 12 Mile Creek site: A reinvestigation. *American Antiquity 49:*757–764.
Rowe, J. H.
1965   An interpretation of radiocarbon measurement on archaeological samples from Peru. In *Proceedings of the Sixth International Conference Radiocarbon and Tritium,* compiled by R. M. Chatters and E. A. Olson. Springfield, Virginia: Clearinghouse for Federal Scientific and Technical Information. Pp. 187–198.
Rubin, M., R. C. Likins, and D. G. Berry
1963   On the validity of radiocarbon dates from snail shells. *Journal of Geology 71:*84–89.
Rubin, M., and M. E. Suess
1955   U. S. Geological Survey radiocarbon dates II. *Science 121:*481–488.
Rubin, M., and D. W. Taylor
1963   Radiocarbon activity of shells from living clams and snails. *Science 141:*637.
Satterthwaite, L., and E. K. Ralph
1960   New radiocarbon dates and the Maya correlation problem. *American Antiquity 26:*165–184.
Saupe, F., O. Strappa, R. Coppens, B. Guillet, and R. Jaegy
1980   A possible source of error in ^{14}C dates: volcanic emanations (examples from the Monte Amiata District, Provinces of Grosseto and Sienna, Italy). *Radiocarbon 22:*525–531.
Säve-Söderbergh, T., and I. U. Olsson
1970   C14 dating and Egyptian chronology. In *Radiocarabon variations and absolute chronology,* edited by I. U. Olsson. Stockholm: Almqvist & Wiksell. Pp. 35–56.
Sayre, E. V., G. Harbottle, R. W. Stoenner, R. L. Otlet, and G. V. Evans
1981   The use of the small gas proportional counters for the carbon 14 measurement of very small samples. *IAEA proceedings on methods of low level counting and spectrometry.* Vienna: International Atomic Energy Agency. P. 393.
Scott, E. M., M. S. Baxter, and T. C. Aitchison
1983 ^{14}C dating reproducibility: Evidence from a combined experimental and statistical programme. In *^{14}C and archaeology,* edited by W. G. Mook and H. T. Waterbolk. Strasbourg: Council of Europe. Pp. 133–145.
1984   A comparison of the treatment of errors in radiocarbon dating calibration methods. *Journal of Archaeological Science* 11:455–466.
Sheppard, J. C.
n.d.   *A radiocarbon dating primer.* Radiocarbon Laboratory, Department of Chemical and Nuclear Engineering, College of Engineering, Washington State University, Pullman.
Sheppard, J. C., J. F. Hooper, and Y. Welter
1983   Radiocarbon dating archaeologic and environmental samples containing 10 to 120 milligrams of carbon. *Radiocarbon 25:*493–500.
Sheppard, J. C., P. E. Wigand, C. E. Gustafson, and M. Rubin
1987   A reevalution of the Marmes rockshelter radiocarbon chronology. *American Antiquity,* 52:118–124.

Siegenthaler, U., M. Heimann, and H. Oeschger
1980   ¹⁴C variations caused by changes in the global carbon cycle. *Radiocarbon* 22:177–191.

Sinex, F. B., and B. Faris
1959   Isolation of gelatin from ancient bones. *Science 129:*969.

Skoog, D. A., and D. M. West
1971   *Principles of instrumental analysis.* New York: Holt, Rinehart and Winston.

Slusher, H. S.
1981   *Critique of radiometric dating.* Technical Monograph No. 2, (2nd Ed.). Institute for Creation Research, San Diego.

Smith, A. G., J. R. Pilcher, and G. W. Pearson
1971   New radiocarbon dates from Ireland. *Antiquity 45:*97–102.

Smith, H. S.
1964   Egypt and C14 dating. *Antiquity 38:*32–37.

Spaulding, A. C.
1958   The significance of differences between radiocarbon dates. *American Antiquity 23:*309–311.

Srdoc, D., N. Horvatincic, B. Obelic, and A. Sliepcevic
1983   Radiocarbon dating of tufa in paleoclimatic studies. *Radiocarbon 25:*421–427.

Srdoc, D., B. Obelic, and N. Horvatincic
1980   Radiocarbon dating of calcareous tufu: How reliable data can we expect? *Radiocarbon 22:*858–862.

Stafford, T. W., Jr., R. C. Duhamel, C. V. Haynes, Jr., and K. Brendel
1982   Isolation of proline and hydroxyproline from fossil bone. *Life Science 31:*931–938.

Stafford, T. W., Jr., A. J. T. Jull, T. H. Zabel, D. J. Donahue, R. C. Duhamel, K. Brendel, C. V. Haynes, Jr., J. L. Bishcoff, L. A. Payen, and R. E. Taylor
1984   Holocene age of the Yuha burial: Direct radiocarbon determinations by accelerator mass spectrometry. *Nature (London) 308:*446–447.

Stenhouse, M. J., and M. S. Baxter
1983   ¹⁴C dating reproducibility: Evidence from routine dating of archaeological samples. In *¹⁴C and archaeology,* edited by W. G. Mook and H. T. Waterbolk. Strasbourg: Council of Europe. Pp. 147–161.

Stephenson, E. J., T. S. Mast, and R. A. Muller
1979   Radiocarbon dating with a cyclotron. *Nuclear Instruments and Methods 158:*571–577.

Stuckenrath, R.
1963   University of Pennsylvania radiocarbon dates VI. *Radiocarbon 5:*82–103.
1965   Carbon-14 and the unwary archeologist. In *Proceedings of the Sixth International Conference Radiocarbon and Tritium Dating,* compiled by R. M. Chatters and E. A. Olson. Springfield, Virginia: Clearinghouse of Federal Scientific and Technical Information. Pp. 304–318.
1977   Radiocarbon: Some notes from Merlin's diary. *Annals of the New York Academy of Sciences 288:*181–188.

Stuiver, M.
1970   Long-term C14 variations. In *Radiocarbon variations and absolute chronology,* edited by I. U. Olsson. Stockholm: Almqvist & Wiksell. Pp. 197–213.
1971   Evidence for the variation of atmospheric ¹⁴C content in the late Quaternary. In *The late Cenozoic glacial ages,* edited by K. K. Turekian. New Haven: Yale University Press. Pp. 69–70.

1978a Carbon-14 dating: A comparison of beta and ion counting. *Science 202*:881–883.

1978b Radiocarbon timescale tested against magnetic and other dating methods. *Nature (London) 273*:271–274.

1980 Workshop on ^{14}C data reporting. *Radiocarbon 22*:964–966.

1982 A high-precision calibration of the AD radiocarbon time scale. *Radiocarbon 24*:1–26.

1983 Calibration of the radiocarbon timescale. In *^{14}C and archaeology,* edited by W. G. Mook and H. T. Waterbolk. [*PACT 8*:498–501]. Strasbourg: Council of Europe. Pp. 498–501.

Stuiver, M., C. H. Heusser, and I. C. Yang

1978 North American glacial history extended to 75,000 years ago. *Science 200*:16–21.

Stuiver, M., and R. Kra (editors)

1980a Proceedings of the Tenth International Radiocarbon Conference—Bern. *Radiocarbon 22*:133–562.

1980b Proceedings of the Tenth International Radiocarbon Conference—Heidelberg. *Radiocarbon 22*:565–1016.

1983 Proceedings of the Eleventh International Radiocarbon Conference—Seattle. *Radiocarbon 25*:171–795.

1986 Proceedings of the Twelfth International Radiocarbon Conference—Trondheim, Norway. *Radiocarbon 28*:177–804.

Stuiver, M., and H. A. Polach

1977 Discussion: Reporting of ^{14}C data. *Radiocarbon 19*:355–363.

Stuiver, M., and P. D. Quay

1980 Changes in atmospheric carbon-14 attributed to a variable sun. *Science 207*:11–19.

1981 Atmospheric ^{14}C changes resulting from fossil fuel CO_2 release and cosmic ray flux variability. *Earth and Planetary Science Letters 53*:349–362.

Stuiver, M., S. W. Robinson, and I. C. Yang

1979 ^{14}C dating to 60,000 years B. P. with proportional counters. In *Radiocarbon dating,* edited by R. Berger and H. E. Suess. Berkeley: University of California Press. Pp. 202–215.

Stuiver, M., and C. S. Smith

1965 Radiocarbon dating of ancient mortar and plaster. In *Proceedings of the Sixth International Conference Radiocarbon and Tritium,* compiled by R. M. Chatters and E. A. Olson. Springfield, Virginia: Clearinghouse for Federal Scientific and Technical Information. Pp. 338–343.

Stuiver, M., and H. E. Suess

1966 On the relationship between radiocarbon dates and true sample ages. *Radiocarbon 8*:534–540.

Suess, H. E.

1954a Natural radiocarbon measurements by acetylene counting. *Science 120*:5–7.

1954b U. S. Geological Survey radiocarbon dates I. *Science 120*:467–473.

1955 Radiocarbon concentration in modern wood. *Science 122*:415–417.

1961 Secular changes in the concentration of atmospheric radiocarbon. In *Problems related to interplanetary matter* (Nuclear Science Series Report Number 33, Publication 845). Washington, D.C.: National Academy of Sciences—National Research Council. Pp. 90–95.

1965 Secular variations of the cosmic-ray produced carbon 14 in the atmosphere and their intrepretations. *Journal of Geophysical Research 70*:5937–5951.

1967 Bristlecone pine calibration of the radiocarbon time scale from 4100 B.C. to 1500 B.C. In *Radiocarbon dating and methods of low-level counting*. Vienna: International Atomic Energy Agency. Pp. 143–150.

1970 Bristlecone-pine calibration of radiocarbon time 5200 B.C. to present. In *Radiocarbon variations and absolute chronology*, edited by I. U. Olsson. Stockholm: Almqvist & Wiksell. Pp. 303–312.

1979 A calibration table for conventional radiocarbon dates. In *Radiocarbon dating*, edited by R. Berger and H. E. Suess. Berkeley: University of California Press. Pp. 777–784.

1980 Radiocarbon geophysics. *Endeavour 4*:113–117.

1986 Secular variations of cosmogenic ^{14}C on earth: their discovery and interpretation. *Radiocarbon 28*:259–266.

Suess, H., and C. Strahm
1970 The neolithic of Auverneir, Switzerland. *Antiquity 44*:91–99.

Swokowski, E. W.
1975 *Calculus with analytic geometry*. Boston: Prindle, Weber and Schmidt.

Tamers, M. A.
1965 Routine carbon-14 dating using liquid scintillation techniques. In *Proceedings of the Sixth International Conference Radiocarbon and Tritium Dating*, compiled by R. M. Chatters and E. A. Olson. Springfield, Virginia: Clearinghouse for Federal Scientific and Technical Information. Pp. 53–67.

1970 Validity of radiocarbon dates on terrestrial snail shells. *American Antiquity 35*:94–100.

Tauber, H.
1970 The Scandinavia varve chronology and C14 dating. In *Radiocarbon variations and absolute chronology*, edited by I. U. Olsson. Stockholm: Almqvist & Wiksell. Pp. 173–196.

1979 ^{14}C activity of arctic marine mammals. In *Radiocarbon dating*, edited by R. Berger and H. E. Suess. Berkeley: University of California Press. Pp. 447–452.

1980 Discussion. *Radiocarbon 22*:198–199.

1983 Possible depletion in ^{14}C in trees growing in calcareous soils. *Radiocarbon 25*:217–420.

Taylor, R. E.
1970 Chronological problems in western Mexican archaeology. A dating systems approach to archaeological research. Unpublished Ph.D. dissertation, University of California, Los Angeles.

1975 UCR radiocarbon dates II. *Radiocarbon 17*:396–406.

1978 Radiocarbon dating: An archaeological perspective. In *Archaeological chemistry II*, edited by G. F. Carter. Advances in Chemistry Series, No. 171. Washington, D.C.: American Chemical Society. Pp. 33–69.

1980 Radiocarbon dating of Pleistocene bone: Towards criteria for the selection of samples. *Radiocarbon 22*:969–979.

1982 Problems in the radiocarbon dating of bone. In *Nuclear and chemical dating techniques: Interpreting the environmental record*, edited by L. A. Currie. Washington, D.C.: American Chemical Society. Pp. 453–473.

1983 Non-concordance of radiocarbon and amino acid racemization deduced age estimates on human bone. *Radiocarbon 25*:647–654.

1985a The beginnings of radiocarbon dating in *American Antiquity*: an historical perspective. *American Antiquity 50*:309–325.

1985b Radiocarbon dating. In *The encyclopedia of physics*, edited by R. M. Basançon (3rd ed.). New York: Van Nostrand-Reinhold. Pp. 1032–1038.

Taylor, R. E., Editor
1981   *Radiocarbon dating in archaeology: Needs and priorities in the 1980s.* A report of a conference held at the National Science Foundation, Washington, D.C., June 10–12.
Taylor, R. E., and R. Berger
1967   Radiocarbon content of marine shells from the Pacific coast of central and south America. *Science 158:*180–1182.
1968   Radiocarbon dating of the organic portion of ceramic and wattle-and-daub house construction materials of low carbon content. *American Antiquity 33:*363–366.
1980   The date of "Noah's Ark." *Antiquity 54:*34–36.
Taylor, R. E., R. Berger and B. Dimsdale
1968   Electronic data processing for radiocarbon dates. *American Antiquity 33:*180–184.
Taylor, R. E., D. J. Donahue, T. H. Zabel, P. E. Damon, and A. T. J. Jull
1984a   Radiocarbon dating by particle accelerators: an archaeological perspective. In *Archaeological chemistry III,* edited by J. B. Lambert. Washington, D.C.: American Chemical Society. Pp. 333–356.
Taylor, R. E., L. A. Payen, and P. J. Slota, Jr.
1984b   Impact of AMS ^{14}C determinations on considerations of the antiquity of *Homo sapiens* in the Western Hemisphere. *Nuclear Instruments and Methods in Physics Research 233*(B5):312–316.
Taylor, R. E. and P. S. Slota, Jr.
1979   Fraction studies on marine shell and bone samples for radiocarbon analysis. In *Radiocarbon dating,* edited by R. Berger and H. E. Suess. Berkeley: University of California Press. Pp. 422–431.
Terasmae, J.
1984   Radiocarbon dating: Some problems and potential developments. In *Quaternary dating methods,* edited by W. C. Mahaney. Amsterdam: Elsevier. Pp. 1–16.
Thomas, D. H.
1978   The awful truth about statistics in archaeology. *American Antiquity 43:*231–244.
Thompson, R.
1983 ^{14}C dating and magnetostratigraphy. *Radiocarbon 25:*299–238.
Throughton, J. H.
1972   Carbon isotope fractionation by plants. In *Proceedings of the 8th International Conference on Radiocarbon Dating,* compiled by T. A. Rafter and T. Grant-Taylor. Wellington: Royal Society of New Zealand. Pp. 420–438.
Tite, M. S.
1972   *Methods of physical examination in archaeology.* New York: Seminar Press.
Topping, J.
1962   *Errors of observation and their treatment* (3rd ed.). London: Chapman & Hall.
Tuross, N., and P. E. Hare
1978   Collagen in fossil bone. *Carnegie Institution of Washington Yearbook 77:*891–895.
Van der Merwe, N. J.
1969   *The carbon-14 dating of iron.* Chicago: University of Chicago Press.
Van der Merwe, N. J., and M. Stuiver
1968   Dating iron by the carbon-14 method. *Current Anthropology 9:*48–53.
Venkatesan, M. I., T. W. Linick, H. E. Suess, and G. Buccellati
1982   Asphalt in carbon-14 dated archaeological samples from Terqa, Syria. *Nature (London) 295:*517–519.

Vita-Finzi, C.
  1980 ^{14}C dating of recent crustal movements in the Persian Gulf and Iranina Makran.
              Radiocarbon 22:763–773.
Vogel, J. C.
  1983 ^{14}C variations during the Upper Pleistocene. *Radiocarbon 25*:213–218.
de Vries, H.
  1957   The removal of radon from CO_2 for use in ^{14}C age measurements. *Applied
              Science Research, Section B 6*:461–470.
  1958   Variations in concentration of radiocarbon with time and location on earth.
              Proceedings, Nederlandsche Akademie van Wetenschappen, Series B 61:1.
de Vries, H., and G. W. Barendsen
  1953   Radiocarbon dating by a proportional counter filled with carbon dioxide. *Physica
              (Amsterdam) 19*:987.
  1954   Measurements of age by the carbon-14 technique [Groningen I]. *Nature (Lon-
              don) 174*:1138–1141.
Walker, A. J., and R. L. Otlet
  1985   Harwell radiocarbon measurements IV. *Radiocarbon 27*:74–94.
Walker, A. J., R. L. Otlet, and A. J. Clark
  1983   Comments on the application of ^{14}C dating to UK archaeology based on ex-
              perience over a number of years of operating service measurements. In *^{14}C
              and archaeology*, edited by W. G. Mook and H. T. Waterbolk. [*PACT 8*:429–
              439]. Strasbourg: Council of Europe. Pp. 429–439.
Wand, J. O.
  1981   Microsample preparation for radiocarbon dating. Unpublished Ph.D. disser-
              tation, Oxford University.
Waterbolk, H. T.
  1960   The 1959 carbon-14 symposium at Groningen. *Antiquity 34*:14–18.
  1971   Working with radiocarbon dates. *Proceedings of the Prehistoric Society 37*:15–
              33.
  1983a   Ten guidelines for the archaeological interpretation of radiocarbon dating. In
              ^{14}C and archaeology, edited by W. G. Mook and H. T. Waterbolk. [*PACT
              8*:57–70]. Strasbourg: Council of Europe. Pp. 57–70.
  1983b   Thirty years of radiocarbon dating: the retrospective view of a Groningen ar-
              chaeologist. In *^{14}C and archaeology*, edited by W. G. Mook and H. T. Wa-
              terbolk. [*PACT 8*:17–27]. Strasbourg: Council of Europe. Pp. 17–27.
  1983c   The integration of radiocarbon dating in archaeology. *Radiocarbon 25*:639–
              644.
Watts, W. A.
  1960   C-14 dating and the neolithic in Ireland. *Antiquity 34*:111–116.
Weinstein, J. M.
  1980   Palestinian radiocarbon dating: a reply to James Mellaart. *Antiquity 54*:21–24.
  1984   Radiocarbon dating in the southern Levant. *Radiocarbon 26*:297–366.
Weisberg, L. H. G., and T. W. Linick
  1983   The question of diffuse secondary growth of palm trees. *Radiocarbon 25*:803–
              809.
Welch, J. J., K. J. Bertsche, P. G. Friedman, D. E. Morris, R. A. Muller, and P. P. Tans
  1984   A 40 keV cyclotron for radioisotope dating. *Nuclear Instruments and Methods
              in Physics Research 233*(B5):230–232.
Wendorf, F., R. Schild, A. E. Close, D. J. Donahua, A. J. T. Jull, T. J. Zabel, H. Wieckowska,
              M. Kobusiewicz, B. Issawi, and N. el Hadidi
  1984   New radiocarbon dates on the cereals from Wadi Kubbankya. *Science 225*:645–
              646.

Wendorf, F., R. Schild, N. el Hadidi, A. E. Close, M. Kobusiewicz, H. Weickowska, B. Issawi, and H. Haas
1979   Use of barley in the Egyptian Late Paleolithic. *Science 205*:1341–1347.
Wigley, T. M. L., and A. B. Muller
1981   Fractionation correction in radiocarbon dating. *Radiocarbon 23*:173–190.
Wilke, P. J.
1978   Cairn burials of the California deserts. *American Antiquity 43*:444–448.
Willey, G. R. and P. Phillips
1958   *Method and theory in American archaeology*. Chicago: University of Chicago Press.
Willey, G. R., and J. A. Sabloff
1980   *A history of American archaeology* (2nd ed). San Francisco: W. H. Freeman and Company.
Willis, E. H.
1969   Radiocarbon dating. In *Science in archaeology, a survey of progress and research*, edited by D. Brothwell and E. Higgs. (2nd ed.). Stolkholm: Almqvist and Wiksell. Pp. 46–57.
Willis, E. H., H. Tauber, and K. O. Munnich
1960   Variations in the atmospheric radiocarbon concentration over the past 1300 years. *Radiocarbon 2*:1–4.
Willkomm, H.
1983   The reliability of archaeologic interpretation of ^{14}C dates. *Radiocarbon 25*:645–646.
Wilson, H. W.
1979   Possibility of measurement of ^{14}C by mass spectrometer techniques. In *Radiocarbon dating*, edited by R. Berger and H. E. Suess. Berkeley: University of California Press. Pp. 238–245.
Winter, J.
1972   Radiocarbon dating by thermoluminescent dosimetry. *Archaeometry 14*:281–286.
Wise, E. N., and D. Shutler, Jr.
1958   University of Arizona radiocarbon dates (I). *Science 127*:72–73.
Wolfli, W., H. A. Polach, and H. H. Anderson (editors)
1984   Accelerator Mass Spectrometry, Proceedings of the Third International Symposium on Accelerator Mass Spectrometry. *Nuclear Instruments and Methods in Physics Research 233*(B5):1–448.
Wolfman, D.
1984   Geomagnetic dating methods in archaeology. In *Advances in archaeological method and theory*, edited by M. Schieffer. (Vol. 7). New York: Academic Press. Pp. 363–458.
De Young, D. B.
1978   Creationist predictions involving C-14 dating. *Creation Research Society Quarterly 15*:14–16.

Illustration Credits

Authors (and publication sources) who have given permission to include previously published figures in this volume—in some cases in a modified form as noted in the captions—include W. F. Libby, University of Chicago Press, Figures 2.3 and 6.2; H. E. Suess, Figure 2.5; E. K. Ralph, MASCA, Figure 2.6; R. M. Clark, *Antiquity*, Figure 2.7; J. Klein, *Radiocarbon*, Figures 2.8, 2.11, and 5.11; G. W. Pearson, *Radiocarbon*, Figure 2.9; W. F. Cain and H. E. Suess, *Journal of Geophysical Research*, copyright by the American Geophysical Union and I. Levin, *Radiocarbon*, Figure 2.14; D. A. Skoog, Holt, Rinehart and Winston, Inc., Figure 4.1; G. Friedlander, reprinted by permission of John Wiley & Sons, copyright 1964, Figure 4.3; H. E. Suess and L. D. Ford, Figures 4.5 and 4.6; R. Berger, Figures 4.8 and 6.2 through 6.7; J. R. Arnold, *Science*, copyright 1954 by AAAS, Figure 4.10; R. Muller, Harwood Academic Publishers, Figure 4.11; T. Jull and G. Kew, copyright 1984 American Chemical Society, Figures 4.12 and 4.13; W. F. Libby, The Royal Society (London), Figure 4.14; H. Willkomm, *Radiocarbon*, 5.1; I. U. Olsson, *Geologiska Foreningen I Stockholm Forhandlingar*, Publishing House of the Swedish Research Councils, Figures 5.5 and 5.6; R. Stuckenrath, New York Academy of Sciences, Figure 5.7; H. Barker, Royal Society (London), Figure 5.8; S. W. Robinson, *Syesis*, British Columbia (Canada) Provincial Museum, Figure 5.9; and M. S. Baxter, *Antiquity*, Figure 5.12.

INDEX